Wesley Stoker Barker Woolhouse

Elements of the Differential Calculus

Sixth Edition

Wesley Stoker Barker Woolhouse

Elements of the Differential Calculus
Sixth Edition

ISBN/EAN: 9783337811266

Printed in Europe, USA, Canada, Australia, Japan

Cover: Foto ©berggeist007 / pixelio.de

More available books at **www.hansebooks.com**

ELEMENTS

OF

THE DIFFERENTIAL CALCULUS

BY

W. S. B. WOOLHOUSE, F.R.A.S., F.S.S.

AUTHOR OF "THE MEASURES, WEIGHTS, AND MONEYS OF ALL NATIONS," ETC.

Sixth Edition, with Index

Capio Lumen

LONDON

LOCKWOOD & CO., 7, STATIONERS' HALL COURT

LUDGATE HILL

1874

PREFACE.

On first commencing to read the Differential Calculus, a subject which opens a wide field of analytical research, the student enters upon an entirely new system of thought. In his previous investigations he has always been accustomed to consider quantities, whether known or unknown, as having some fixed or determinate value; he has now to conceive the values of certain quantities to undergo continuous changes, and to operate upon these changes with new symbols and new processes, which in themselves have but a remote analogy to ordinary Algebra.

When two quantities, thus continuously variable, are connected by an analytical equation, and their values are therefore mutually dependent on each other, and they are supposed to be affected by simultaneous changes, it is evident that the increments will also be connected by some corresponding analytical relation. The primary object of the Calculus is to establish general methods of investigating the nature and properties of such relations when the changes or increments are supposed to be small. To effect this, it is first requisite to trace the successive values of the ratio subsisting between two increments, when the increments themselves are supposed to continuously decrease in magnitude, and to determine the limiting value of this ratio when they ultimately become infinitesimals. This ultimate or limiting value is, in fact, that which represents the ratio $\frac{0}{0}$ when the increments are supposed absolutely to vanish, and it is completely defined and accurately determined by referring the successive values to the recognized law of continuity. The operation here described is the true foundation of the Calculus, and the condition of continuity, especially insisted upon in the present treatise, entirely removes from the limiting value that obscure and indeterminate character which otherwise forms an insuperable obstacle to a proper comprehension of the first principles.

We recommend the student to make himself familiar with the methods of " limiting ratios " and " infinitesimals." The theory of Infinitesimals is literally that of the Differential Calculus, and the principal law which regulates this theory is directly inferred from the method of limiting ratios. The two methods are indeed virtually but modifications of the same idea.

Thus, in comparing together the relative values of any two infinitesimals, the rejection of terms involving infinitesimals of higher orders is, in effect, precisely the same as that of proceeding to the ultimate ratio of the infinitesimal quantities by the method of limits, and such rejection may in reality be said to be the operation of cropping down the quantities to their ultimate or limiting relative proportions. The method of infinitesimals, sometimes called the method of elements, is therefore as correct in its reasonings and deductions, and as accurate in its results, as the method of limits, and, being less abstract in its nature, its application, when properly understood, is usually attended by greater facility and clearness, especially in abstruse investigations.

In preparing the present publication, we have endeavoured to do justice to each Chapter by restricting the applications to matters of general interest, which was considered to be essentially more solid and satisfactory than any attempt to give, within the prescribed limits, a meagre outline of a more extended variety of subjects. The first five Chapters comprise the entire theory of the Calculus as a pure branch of analysis, and the remaining Chapters exhibit the applications to the theory of maxima and minima, and the geometry of curve lines. The general theorems of Euler, Lagrange, and Laplace not being essentially required in the body of the work, though very important to be known by those who may desire to extend their course of reading, are inserted at the end of the last Chapter.

The subjects contained in the several Chapters are treated according to the most elegant and approved methods of investigation, some of which are presumed to be new; numerous interesting examples, exhibiting their respective results, are inserted for the exercise of the student, and copious explanations are given of the precise nature of the principles involved in the various operations. It is hoped that these explanations may tend to obviate the peculiar difficulties so commonly experienced in the acquirement of correct notions, and, by making good the foundation, conduce to the rational and satisfactory advancement of the intelligent student in obtaining a knowledge of one of the greatest superstructures of the human intellect. Should this expectation be in any degree realized, we shall experience a corresponding gratification.

CONTENTS.

Chapter I. — Definitions and First Principles.

Chapter II.—Differentiation of Functions.

Chapter VIII.—Formulæ for Polar Equations, &c.

THE DIFFERENTIAL CALCULUS.

CHAPTER I.

DEFINITIONS AND FIRST PRINCIPLES.

(1.) BY means of Algebra we investigate the various numerical and symbolical relations subsisting amongst fixed quantities, some of which are known and others unknown, the ultimate object in general being to evolve the unknown values, or to express them in terms of those which are known.

In the Differential Calculus certain values or quantities related to each other are supposed to continuously increase or decrease in value, and our object is to investigate the relations subsisting amongst the corresponding changes that take place in their values when those changes are indefinitely diminished. Although the changes themselves are supposed to be infinitely small, it will be found that the ratios which these changes bear to one another are usually finite and appreciable, and therefore suitable subjects of investigation.

(2.) The symbols which enter into the operations of the Differential Calculus are of two kinds, representing *constant* quantities and *variable* quantities.

A *constant* quantity is one which retains the same determinate value, this value being unaffected by the supposed changes in other quantities.

A *variable* quantity is one which admits of a succession of different values.

(3.) A variable quantity varies *continuously* when in changing

▲

from one value to another it passes through every intermediate value. For example, if a point be supposed to move along a curve line it will do so continuously, since in moving from one position to another it must have passed through every intermediate point. It follows therefore that quantities which vary continuously may be supposed to increase or decrease by very small variations, capable of being diminished to any extent.

(4.) A *function* is any analytical expression involving one or more variable quantities, and is usually called a function of the variable quantity or quantities which it contains. Thus x^2, $x^2 + ax$, $\sqrt{a^2 - x^2}$ are functions of x, and $ax + by$, $x^2 + y^2 + xy$ are functions of x and y.

Functions are frequently denoted by prefixing one of the characters F, f, ϕ, ψ, &c. to the variable or variables, and for brevity they are sometimes indicated by a single letter.

Functions are the same in form when the quantities are involved in the same manner. Thus $x^2 + ax$ is the same function of x that $y^2 + ay$ is of y; and supposing F to be the characteristic of $x^2 + ax$, that is, supposing $x^2 + ax$ to be indicated by Fx, the expression $y^2 + ay$ will be similarly indicated by Fy. In like manner if $x^2 + y^2 + xy$ be represented by $f(x, y)$, the expression $u^2 + v^2 + uv$ would be denoted by $f(u, v)$.

Functions which, in a finite number of terms, involve the ordinary algebraical operations of addition, subtraction, multiplication, division, involution and evolution, are called *Algebraical Functions*. According to this definition, $ax + b$,

$$a^2 - x^2, \frac{a^2 + bx}{b^2 - x^2}, (a - x)\sqrt{b^2 + x^2}, \frac{b - x}{b + x}(a^2 - bx + x^2)^{\frac{3}{2}}$$

and all expressions belonging to pure Algebra, are algebraical functions.

Functions which do not exhibit the ordinary algebraical operations and which do not admit of being so expressed in finite terms, are called *Transcendental Functions*. Thus a^x, log x, sin x, are transcendental functions; the first being exponential, the second logarithmic, and the third trigonometrical. There are other transcendental functions besides these, arising out of certain special researches, but it will not be necessary to particularize any of them here.

(5.) When a variable quantity x is assumed to pass to another value, the amount of change or the difference between the two values is called an *Increment* or *Difference*. Similarly the difference between the two corresponding values or the corresponding change that takes place in the value of any function of x is the increment or difference of the function. These increments are usually denoted by prefixing the symbol Δ. Thus Δx, $\Delta(fx)$ are simultaneous increments of x and fx, the corresponding new values being $x + \Delta x$ and $f(x + \Delta x)$ or $fx + \Delta(fx)$. When a value becomes decreased by the supposed change, the increment is to be understood as having a negative value.

(6.) Let $u = fx$ denote a function of a variable quantity x. Suppose x to receive a small increment Δx so as to become of the value $x + \Delta x$, and let the corresponding value of u be supposed to be $u + \Delta u = f(x + \Delta x)$. Let the binomial function $f(x + \Delta x)$, when expanded in terms involving the integral powers of Δx, be also supposed to give

$$u + \Delta u = f(x + \Delta x) = fx + P\Delta x + Q\Delta x^2$$
$$+ R\Delta x^3 + \&c. \ldots (1)$$

in which P, Q, R, &c. are new functions of x, independent of Δx, and owing their forms entirely to that of fx; also Δx is to be regarded as a single symbol, so that Δx^2, Δx^3, &c. indicate $(\Delta x)^2$, $(\Delta x)^3$, &c. From this and the initial equation $u = fx$, we deduce

$$\Delta u = P\Delta x + Q\Delta x^2 + R\Delta x^3 + \&c. \ldots (2)$$

and this value would represent the difference or increment of the function u according to the theory of Finite Differences.

We have also, dividing by Δx,

$$\frac{\Delta u}{\Delta x} = P + Q\Delta x + R\Delta x^2 + \&c. \ldots (3)$$

Each step in this deduction, including the division by Δx, is free from ambiguity when Δx is of any value, great or small, positive or negative; but the result has no intelligible signification when Δx is zero, for as soon as Δx absolutely vanishes, we immediately lose all idea of quantity on the left-hand side of the equation, and the fraction takes the singular and indeterminate form $\dfrac{0}{0}$. As, however, the equation must obviously hold for every other value excepting $\Delta x = 0$, we may take Δx extremely small, and it still will be strictly true for every value between that and zero; and as there is no symbol of discontinuity on the right-hand side of the equation, we may, by applying the principle of continuity to the fraction, include the existence of the equation, when Δx actually vanishes. Thus we should have

$$\frac{\Delta u}{\Delta x} \text{ (when } \Delta x = 0) = \frac{0}{0} = P \ . \ . \ . \ . \ (4)$$

and the coefficient P will therefore represent the limiting value of the fraction $\dfrac{\Delta u}{\Delta x}$, when Δu and Δx simultaneously vanish; and here we must not overlook the implied condition that the particular value thus assigned to the vanishing fraction when it reaches its indeterminate state $\dfrac{0}{0}$, is determined by a consideration of its successive values and is that which obeys the continuity existing amongst all the other values as Δx continuously diminishes from a small position to a small negative value. This condition of continuity forms the basis of what is usually called the "theory of limits" or of "limiting ratios," and should be well understood by the student, who will afterwards not experience any difficulty in acquiring a true conception of the first principles and objects of the Calculus.

The equation (3) has been made to merge into the equation (4) by supposing the increments Δu and Δx to absolutely vanish. It is evident that the former equation will assimilate to the latter to any degree of nearness by conceiving the values of $\wedge u$, Δx to diminish, and that they will be indefinitely near when Δx is indefinitely small. In order therefore to impart some tangible signification to the symbols on the left-hand side of the equation (4), the values of Δu, Δx, instead of being absolute zeros, are supposed to be extremely small quantities having the same ratio to each other as the limiting ratio expressed by the equation, and they are then designated by du, dx. The equation is therefore stated as follows :

$$\left. \begin{aligned} \frac{du}{dx} &= P \\ \text{or } du &= P\,dx \end{aligned} \right\} \quad \dots \dots \text{(5)}$$

The indefinitely small quantities du, dx, thus related, are called the *differentials* of u and x, so that $P\,dx$ represents the value of the differential of the function u; and from what has preceded it is evident that the smaller dx is conceived to be as a change in the value of x, the more nearly will du assimilate to the actual corresponding change in the value of u.

The quantity x which is first supposed to vary and on the differential of which other differentials are thus made to depend is called the *independent variable.*

The coefficient P is called the *differential coefficient* of the function u, *with respect to x*, because it is the coefficient or multiplier of the differential dx which determines the differential of the function.

The student will observe that in the Calculus the letter d is not in any case employed as it may be in Algebra, to represent quantity or value. In this sense it has no isolated signification, and it is never used excepting as the symbol of operation which characterizes the differential of the variable to which it is immediately prefixed.

(7.) The peculiar difficulty in the preceding deductions is precisely analogous to that which occurs in conveying an adequate idea of the measurement of the velocity of a body when that velocity is continuously variable. When the velocity is uniform, the space and time will vary proportionally, and the velocity will be correctly represented by the ratio, or fraction,

$$\frac{\text{space described}}{\text{time of describing it}}$$

which ratio, or fraction, will preserve the same value whether the space and corresponding time be taken great or small. But when the velocity is variable it is obvious that the above fraction cannot accurately define its value at the point from which the space is supposed to be measured, because the space, however small, will then be described by a continuous succession of different velocities. It is however evident that the smaller and smaller the space and time are taken, the closer will their ratio approximate to the true velocity, and that the diminishing error of such approximation will become completely exhausted when we take the limiting ratio as the quantities are supposed to vanish. The velocity of the body at any point is therefore represented with rigorous exactness by the limiting value of the above fraction when it takes the form $\frac{0}{0}$. And thus by analogy the differential coefficient of any function might be defined to be the velocity with which it increases when the independent variable varies uniformly at a rate, to be taken as the unit of measurement. In the geometrical application of this idea, which was the origin of Sir Isaac Newton's method of fluxions, a line is supposed to be generated by the motion, or flowing, of a point, a surface is supposed to be generated by the motion of a line, and a solid by the motion of a surface. It should be observed however that our preconceived notions as to the estimation of velocities of movement, though serving the purpose of illustration, are not sufficiently elementary to be made the basis of a branch of pure science.

The particular considerations under which the equation (2) has been converted into the differential equation (5) conduct us to the ingenious theory propounded by Leibnitz, called the theory of infinitesimals, the principles of which may now be briefly explained.

(8.) Before entering upon this part of the subject it should first be premised that the phrases "infinite number" and "infinitely small quantity," which embody the principal objects of our reasonings, are to be understood as having only a relative signification, since all operations connected with them in the literal or absolute sense of the terms are inconceivable. Thus an "infinite number" is to be considered in a qualified sense as infinitely great in comparison with any finite number; and an "infinitely small quantity" is also to be relatively considered as infinitely small in comparison with any finite quantity.

If any finite quantity be supposed to be divided into an infinite number of parts, each part will be infinitely small and is called an *infinitesimal*, because an infinite number of these is required to make up the finite quantity; it is also when compared with other infinitesimals said to be of the first order. By supposing one of these infinitesimals to be similarly divided into an infinite number of smaller parts, each of these is called an infinitesimal of the second order, and an infinite number of them will be required to make up an infinitesimal of the first order. In like manner by supposing each successive infinitesimal to be divided into an infinite number of parts, infinitesimals of still higher orders are obtained.

The same process also leads us to the conception of different orders of infinities, the word infinity, as before, having only a relative and qualified signification. Thus the number of infinitesimals of the first order contained in the finite quantity, viz. the infinite number of parts into which it is divided, is an infinity of the first order; the number of infinitesimals of the second order contained in the finite quantity is an infinity of the second order, &c., &c. It is evident therefore that infinitesimals and infinities, of the same order, are reciprocally related, since the one multiplied by the other produces the finite quantity. Sometimes an infinitesimal is called an "element" of the integral or finite quantity of which it forms a part.

Referring to the equation (2) in which x, as usual, is supposed to represent an arithmetical value, we may assume $\Delta x = \dfrac{1}{N}$, N denoting any number or numerical value. When N is a large number, Δx becomes a small quantity, and a term $P \Delta x$ which involves its first power is in such case usually called a small quantity of the first order with respect to Δx; $Q \Delta x^2$ which involves the second power is of a still smaller scale of value, and is said to be of the second order with respect to Δx; $R \Delta x^3$ is called a small quantity of the third order with respect to Δx, &c. If N be supposed to be an infinite number, Δx will become an infinitesimal, and denoting it by dx, we have

$$P\,dx = \frac{P}{N}$$

$$Q\,dx^2 = \frac{Q}{N^2} = \frac{Q\,dx}{N}$$

$$R\,dx^3 = \frac{R}{N^3} = \frac{R\,dx^2}{N}$$

&c.   &c.

Hence as P, Q, R, &c. are supposed to be finite coefficients, it follows, according to the preceding definitions, that the terms $P\,dx$, $Q\,dx^2$, $R\,dx^3$, &c. are infinitesimals severally of the first, second, third, &c. orders.

By supposing the number of parts into which the finite quantity is divided to be progressively augmented, the corresponding infinitesimal will become diminished, and in the extreme case the quantity may be assumed to be divided into an infinite number of parts, in the absolute sense of the term, in which case it is easy to conclude that each of the parts must become ultimately zero. In thus proceeding to the extreme case, the nature of the reasoning is in effect the same as that employed in deducing the limiting ratio or ultimate value of a vanishing fraction. The laws of infinitesimals are also founded upon this extreme case, and their operation is always exact, for this simple reason, that the extreme limit $dx = 0$ is, in all mathematical investigations, understood to be applied to the final result of infinitesimal deductions. These laws are as follows :

I. In any equation containing terms of finite value, other terms which represent infinitesimal quantities may be omitted; because in the extreme case these infinitesimals become absolute zeros.

Thus in equation (3) when Δx, Δu become infinitesimals denoted by dx, du, the fraction $\dfrac{du}{dx}$ being not necessarily an infinitesimal, the equation, according to this rule, becomes

$$\frac{du}{dx} = \text{P},$$

being in fact the same as the extreme limit of the equation before expressed in (4) or (5).

II. In an equation containing infinitesimal quantities of any order, all infinitesimals of higher orders may be omitted.

For example, in the equation (2) if Δx become an infinitesimal dx, the terms du, $\text{P}\,dx$ will be infinitesimals of the first order, and the other terms will be infinitesimals of higher orders. Therefore, omitting these, the equation will become

$$du = \text{P}\,dx.$$

This evidently follows by first deducing the equation (3) and then taking the extreme limit as before.

III. In comparing two infinitesimal quantities, if they are of the same order they will have a finite ratio to each other, but if of different orders the ratio will be either zero or infinity.

For example, let $\text{A}\,dx^m$, $\text{B}\,dx^m$ be two infinitesimals, both of the mth order with respect to dx, then

$$\frac{\text{A}\,dx^m}{\text{B}\,dx^m} = \frac{\text{A}}{\text{B}}, \text{ a finite ratio.}$$

Again, let $\text{A}\,dx^{m+n}$, $\text{B}\,dx^m$ be two infinitesimals of the

$(m+n)$th and mth orders respectively, then

$$\frac{A\,dx^{m+n}}{B\,dx^{m}} = \frac{A\,dx^{n}}{B},\text{ an infinitesimal of the } n\text{th order,}$$

$$\frac{B\,dx^{m}}{A\,dx^{m+n}} = \frac{B}{A\,dx^{n}},\text{ an infinity of the } n\text{th order;}$$

and, at the extreme limit, these become

$$\frac{A\,dx^{m+n}}{B\,dx^{m}} = 0 \qquad\qquad \frac{B\,dx^{m}}{A\,dx^{m+n}} = \infty.$$

(9.) The method of determining the position of a tangent to a plane curve supplies an elegant geometrical elucidation of the signification of the differential co-efficient of a function. Let A P B be a curve line; P a point in the curve the coordinates of which are $AD = x$, $DP = y$; Q another point in the curve the coordinates of which are $AD' = x$ $+ \Delta x$, $D'Q = y + \Delta y$; and suppose the curve to be determined by an equation of the form $y = fx$, any function of x.

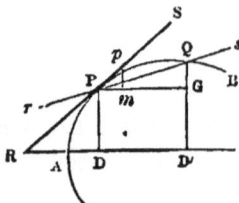

Then from what precedes,

$$\Delta y = P\,\Delta x + Q\,\Delta x^2 + R\,\Delta x^3 + \&c.$$

$$\frac{\Delta y}{\Delta x} = P + Q\,\Delta x + R\,\Delta x^2 + \&c.$$

In the diagram, $\Delta x = PG$, $\Delta y = GQ$, and therefore $\frac{\Delta y}{\Delta x} = \tan \angle s\,PG$. Consequently

$$\tan \angle s\,PG = P + Q\,\Delta x + R\,\Delta x^2 + \&c. \ldots \ldots (a)$$

From this equation we infer that if Δx be taken less and less towards zero, the value of $\tan s\,PG$ will approximate to the differential coefficient (P) as its utmost limit. For the geometrical limit of the angle $s\,PG$, as Δx decreases, we may suppose the point Q to approach nearer and nearer to the point P, and watch the progress of the line rs which passes through them, or we may suppose the line rs to turn gradually about the fixed point P, so that the intersection Q shall proceed towards P. The former of these suppositions will lead ultimately to an indeterminate result, whilst the latter will proceed at once to the extreme limit. Thus on the former supposition, when the point Q finally arrives at the point P, and the two points become one, it is evident that an indefinite number of lines can be drawn through them, and therefore that the position of the line rs is so far indeterminate. But on the other supposition, if the motion of rs be conceived to cease the instant the point Q arrives at the point P, it will then assume the position of the tangent R S, which touches the curve at the point P; and this is obviously the only position which can obey the law of continuity amongst the positions that precede it. If we now suppose the motion of rs to continue onward, it is evident that it will begin to intersect the curve on the other side of the point P, or between P and A, and that the positions will then have reference to negative values of Δx. The line rs will thus pass through a continuous series of positions as Δx gradually diminishes from positive to negative values; and when $\Delta x = 0$, though the position, as depending on the two points through which it has to pass, is then indeterminate, yet the position R S is the only one that can partake of the continuity existing amongst all the others, and the angle S P G is the only one that can partake of the continuity existing amongst the preceding and following values of that angle. Now, according to the equation (α), the series

$$P + Q \Delta x + R \Delta x^2 + \&c.$$

strictly corresponds with the value of tan s P G for every value of Δx except *zero;* and hence as the values of this series as Δx passes from positive to negative values are wholly continuous, and consequently, when $\Delta x = 0$, the first term P partakes of that continuity, it is conclusive that

$$\tan \text{SPG} = P = \frac{dy}{dx} \ \ . \ . \ . \ . \ (\beta)$$

which may be either considered as a fraction whose numerator
and denominator are the differentials of the ordinates, or the
differential coefficient of y considered as a function of x.

By this result it is evident that the differentials of the ordinates x, y may be relatively conceived as represented by two
small coordinate lines Pm, mp terminating in the tangent at
a contiguous point p.

(10.) After what has now been explained the student will
not fail to observe that the leading principle of the Calculus
arises out of the following considerations :

When a fraction, which in a particular case takes the indeterminate form $\frac{0}{0}$, expresses the value of a quantity which we
have reason to know from the nature of the subject does not
become discontinuous in that case, or generally when such a
fraction enters in any equation, the other terms of which are
not discontinuous, the fraction is, under such circumstances,
necessarily limited to continuous values, and consequently,
when the numerator and denominator vanish, it must take the
particular limiting value assigned by the law of continuity. It
is on the ground of continuity alone that the mathematical
accuracy and logical rigour of the principles and applications of
the Calculus may be considered to rest. The fundamental
principle of our operations, according to the theory of limits,
consists in this, that if the increment of a function be divided
by the corresponding increment of the independent variable,
then as the increments are taken less and less towards zero, so
will the quotient approximate in value to the differential coefficient as its utmost limit. Thus the differential coefficient
is that particular value of the vanishing fraction which con-
forms to the law of continuity amongst the other values : and
since this is the identical value of the fraction, which always
enters as the subject of investigation, the truth of the principle
on which the Calculus is applied, in the case of limits, may be
regarded in the strictest sense, and at the same time rendered
clear and satisfactory to the understanding.

(11.) There is yet another mode of laying down the first principles of the Calculus, which, at the onset, has the advantage of obviating all considerations of infinitesimals and limiting ratios, so as to bring the subject within the scope of ordinary Algebra. This method, commonly called " the method of derived functions," is presented by Lagrange in his 'Théorie des Fonctions Analytiques,' and the investigations, which in their nature are purely algebraical, are at the same time elegant, systematic and logical. In substance this method is equivalent to the following :

Let h denote a small accession to the value of a variable quantity x which thereby becomes of the value $x + h$; and suppose the binomial function $f(x + h)$, when developed according to the powers of h, to be as in equation (1), viz. :

$$f(x + h) = fx + P h + Q h^2 + R h^3 + \&c.$$

in which P, Q, R, &c., as before, denote new functions of x whose forms depend wholly upon that of fx.

Then the coefficient P, which is identical with the differential coefficient, Lagrange defines to be the first derived function; he designates it by $f'x$, and observes that it is quite independent of the value of h. By treating the derived function $f'x$ in the same manner, that is, by expanding $f'(x + h)$ and again taking the coefficient of h, a second derived function, designated by $f''x$, is obtained; and this process is further supposed to be successively repeated to third, fourth, &c. derived functions.

(12.) These definitions being premised, the more immediate objects of the calculus of derived functions are :

1. The form of any function fx being given, to determine the forms of the derived functions, and to effect generally the form of the development of the binomial function $f(x + h)$, with other problems relating to the expansion of functions.

2. The form of a derived function being given, to find that of the original or primitive function, &c., &c.

The problems comprised in the first of these are equivalent to those of the Differential Calculus; and those of the second, which refer to the inverse operations of the Calculus, are in

effect the same as the inverse processes of integrating differentials and differential equations in the Integral Calculus.   And
these abstract analytical problems, which embody the essential
principles of the Calculus as an instrument of investigation, are
thus established without introducing any ideas relating to
infinitely small quantities or limiting ratios, all considerations
of small quantities being in fact deferred to their legitimate and
inevitable occurrence when we come to the actual applications
of the Calculus to the various geometrical and physical subjects
which arise in the different branches of mathematical science.

We have here given a brief exposition of the fundamental
principles according to different methods of treatment, because
a knowledge of each of these will be necessary to enable the
student eventually to acquire a thorough command of the
powerful resources of the Calculus.   After a little experience
he will not fail to discover that the collective reasonings employed in these methods are substantially alike, and that they
in reality constitute the same grand unique system of deduction,
only exhibited under different points of view or modified for
the purpose of more immediate adaptation to particular objects
of investigation.

(13.) Before entering upon the manual operations of the
Calculus or discussing the practical methods of differentiating
functions, we shall here concisely repeat those preliminary
ideas respecting the operation of differentiation, which should
in the first place be distinctly impressed upon the mind :

If, when the variable quantity x increases by an increment
Δx, a function u or fx increases by Δu or $\Delta (fx)$; then the
"differential coefficient" of the function is determined by
ascertaining the ultimate ratio of the increments, or the limiting
continuous value of the fraction $\dfrac{\text{increment function}}{\text{increment variable}} = \dfrac{\Delta u}{\Delta x}$ or
$\dfrac{\Delta (fx)}{\Delta x}$ when the increments are supposed to vanish, and this
differential coefficient is symbolized by $\dfrac{du}{dx}$ or $\dfrac{d (fx)}{dx}$, and
sometimes more briefly by u' or $f'x$.

If we further suppose the expansion of the binomial function $f(x + \Delta x)$, according to the ascending powers of Δx, to be

$$f(x + \Delta x) = fx + \mathrm{P}\,\Delta x + \mathrm{Q}\,\Delta x^2 + \&\text{c.};$$

then the coefficient P of Δx, exhibited by the second term, will also be the differential coefficient of the function $f(x)$; that is,

$$\frac{du}{dx} \text{ or } \frac{d(fx)}{dx} = \mathrm{P}.$$

In these relations du and dx may be regarded as simultaneous infinitesimal increments of u and x; but this idea is not always necessary, because $\frac{du}{dx}$ may be either considered as a fraction determining the ultimate ratio of two infinitesimals or as an abstract symbolical representation of the coefficient P, according to the nature of the investigation.

The following examples, in which the differentials are determined from first principles, will practically explain their operation.

Example 1.—Let $u = x^2$; then, as the equation is general for all values of x, when x becomes $x + \Delta x$ it will give

$$(u + \Delta u) = (x + \Delta x)^2 = x^2 + 2x\,\Delta x + \Delta x^2.$$

From this take away the first value $u = x^2$, and we get

$$\Delta u = 2x\,\Delta x + \Delta x^2 \quad \therefore \quad \frac{\Delta u}{\Delta x} = 2x + \Delta x.$$

This last equation is accurately true for all values of Δx, however small, and the value of $2x + \Delta x$ on the right-hand side, will evidently change continuously as we suppose Δx to continuously diminish and ultimately to vanish. Hence making $\Delta x = 0$ and taking the limiting value of the fraction $\frac{\Delta u}{\Delta x}$, denoted by $\frac{du}{dx}$, we obtain

$$\frac{du}{dx} = 2x \qquad \text{or } du = 2x\,dx,$$

which is the differential of the proposed function $u = x^2$.

Example 2.—Let $u = x^3 + 3a^2x$; then, when x becomes $x + \Delta x$,

$$u + \Delta u = (x + \Delta x)^3 + 3a^2(x + \Delta x)$$
$$= x^3 + 3a^2x + 3(x^2 + a^2)\Delta x + 3x\Delta x^2 + \Delta x^3.$$

Reject $u = x^3 + 3a^2x$, and

$$\Delta u = 3(x^2 + a^2)\Delta x + 3x\Delta x^2 + \Delta x^3$$

$$\therefore \frac{\Delta u}{\Delta x} = 3(x^2 + a^2) + 3x\Delta x + \Delta x^2.$$

Hence, as before, making $\Delta x = 0$ and taking the limit, we get

$$\frac{du}{dx} = 3(x^2 + a^2) \qquad \text{or } du = 3(x^2 + a^2)\,dx.$$

Example 3.—Let $u = \dfrac{a^2 + bx}{b - x}$;

then $u + \Delta u = \dfrac{a^2 + b(x + \Delta x)}{b - x - \Delta x}$, and

$$\Delta u = \frac{a^2 + bx + b\Delta x}{b - x - \Delta x} - \frac{a^2 + bx}{b - x} = \frac{(a^2 + b^2)\Delta x}{(b - x)(b - x - \Delta x)}$$

$$\therefore \frac{\Delta u}{\Delta x} = \frac{a^2 + b^2}{(b - x)(b - x - \Delta x)}.$$

Therefore, at the limit,

$$\frac{du}{dx} = \frac{a^2 + b^2}{(b - x)^2} \quad \text{or } du = \frac{a^2 + b^2}{(b - x)^2}\,dx.$$

The process of finding the differential coefficient or the differential of any proposed function is called "differentiation," and we proceed in the following Chapters to establish the principal rules by which we are guided for the purpose of facilitating the actual performance of this operation on the different forms and varieties of functions.

CHAPTER II.

DIFFERENTIATION OF FUNCTIONS.

1. *Algebraical Functions.*

(14.) A constant quantity connected with a function by the sign of addition or subtraction will disappear after differentiation.

Let $u = P \pm c$, P denoting any function of a variable x. When x becomes $x + \Delta x$, suppose P and u to respectively become $P + \Delta P$, $u + \Delta u$; then

$$u + \Delta u = (P + \Delta P) \pm c.$$

From this subtract $u = P \pm c$ and there remains the increment $\Delta u = \Delta P$. Therefore $\dfrac{\Delta u}{\Delta x} = \dfrac{\Delta P}{\Delta x}$ and hence $\dfrac{du}{dx} = \dfrac{dP}{dx}$ or $du = dP$, in which result the constant quantity c does not appear.

(15.) A constant quantity connected with a function as a multiplier or divisor will remain as a multiplier or divisor after differentiation.

Let $u = cP$, P as before denoting any function of a variable x; then when u, P take the new values $u + \Delta u$, $P + \Delta P$, we have

$$u + \Delta u = c (P + \Delta P).$$

From this subtract $u = cP$, and we get $\Delta u = c \Delta P$

$$\therefore \frac{\Delta u}{\Delta x} = c \frac{\Delta P}{\Delta x}.$$

Hence $\dfrac{du}{dx} = c \dfrac{dP}{dx}$ or $du = c\, dP$.

Similarly, if $u = \dfrac{P}{c}$, we find $\dfrac{du}{dx} = \dfrac{1}{c} \cdot \dfrac{dP}{dx}$ or $du = \dfrac{dP}{c}$.

(16.) The differential of a function consisting of two or more terms, connected by the signs of addition or subtraction, is found by differentiating each term separately and collecting the results with their proper signs.

Let $u = P \pm Q \pm R \pm$ &c., where P, Q, R, &c. are functions of x; then when x takes the value $x + \Delta x$, the function u will become

$$u + \Delta u = (P + \Delta P) \pm (Q + \Delta Q) \pm (R + \Delta R) \pm \text{&c.}$$

From this subtracting the former value $u = P \pm Q \pm R \pm$ &c., we get

$$\Delta u = \Delta P \pm \Delta Q \pm \Delta R \pm \text{&c.}$$

$$\therefore \frac{\Delta u}{\Delta x} = \frac{\Delta P}{\Delta x} \pm \frac{\Delta Q}{\Delta x} \pm \frac{\Delta R}{\Delta x} \pm \text{&c.}$$

$$\text{Hence } \frac{du}{dx} = \frac{dP}{dx} \pm \frac{dQ}{dx} \pm \frac{dR}{dx} \pm \text{&c.}$$

$$\text{or } du = dP \pm dQ \pm dR \pm \text{&c.}$$

(17.) The differential coefficient of any constant power of the independent variable x is found by multiplying by the exponent and diminishing the exponent by unity.

Let $u = x^n$; then when x takes the value $x + \Delta x$, $u + \Delta u = (x + \Delta x)^n$.

$$\therefore \Delta u = (x + \Delta x)^n - x^n.$$

To find the value of Δu in powers of Δx it will be necessary to expand this binomial; but the second term of this expansion will suffice for our present object, and this may be readily found by means of induction, independently of the binomial theorem.

First, suppose the exponent n to be a positive integer. By multiplying successively by $x + \Delta x$, disregarding the terms which involve the second and higher powers of Δx, and indicating those terms by $+$ &c., we obtain

$$(x + \Delta x) \ = x \ + \Delta x$$

$$(x + \Delta x)^2 = x^2 + 2x\Delta x + \text{&c.}$$

$$(x + \Delta x)^3 = x^3 + 3x^2 \Delta x + \&c.$$

$$(x + \Delta x)^4 = x^4 + 4x^3 \Delta x + \&c.$$

&c. &c.

And generally, $(x + \Delta x)^n = x^n + n x^{n-1} \Delta x + \&c.$
The value of Δu is therefore of the form

$$\Delta u = n x^{n-1} \Delta x + Q \Delta x^2 + R \Delta x^3 + \&c.$$

where Q, R, &c. denote certain functions of x and n. Hence

$$\frac{\Delta u}{\Delta x} = n x^{n-1} + Q \Delta x + R \Delta x^2 + \&c.;$$

and this equation is true for all values of Δx. By proceeding continuously to $\Delta x = 0$ and taking the limiting value of the fraction, it ultimately gives

$$\frac{du}{dx} = n x^{n-1} \text{ or } du = n x^{n-1} dx.$$

The same reasoning and the same result also obtain when x instead of being considered the independent variable is supposed to represent any function of another variable.

Secondly, suppose the exponent to be a negative integer, or $u = x^{-n}$; then $u = \dfrac{1}{x^n}$, $u + \Delta u = \dfrac{1}{(x + \Delta x)^n}$ and

$$\Delta u = \frac{1}{(x + \Delta x)^n} - \frac{1}{x^n} = -\frac{(x + \Delta x)^n - x^n}{x^n (x + \Delta x)^n}$$

$$= -\frac{n x^{n-1} \Delta x + Q \Delta x^2 + R \Delta x^3 + \&c.}{x^n (x + \Delta x)^n}$$

$$\therefore \frac{\Delta u}{\Delta x} = -\frac{n x^{n-1} + Q \Delta x + R \Delta x^2 + \&c.}{x^n (x + \Delta x)^n}$$

By proceeding as before to the limiting value, this gives

$$\frac{du}{dx} = -\frac{n x^{n-1}}{x^{2n}} = -n x^{-n-1} \text{ or } du = -n x^{-n-1} dx.$$

Thirdly, suppose the exponent to be fractional, or $u =$

$x^{\frac{m}{n}}$; then $u^n = x^m$ and $n\,u^{n-1}\,du = m\,x^{m-1}\,dx$

$$\therefore \frac{du}{dx} = \frac{m\,x^{m-1}}{n\,u^{n-1}} = \frac{m\,x^{m-1}}{n\,x^{\frac{m}{n}(n-1)}} = \frac{m}{n}\,x^{\frac{m}{n}-1}.$$

If the fractional exponent be negative, or $u = x^{-\frac{m}{n}}$; then u^n $= x^{-m}$ and $n\,u^{n-1}\,du = -m\,x^{-m-1}\,dx$, which in the same

way gives $\dfrac{du}{dx} = -\dfrac{m}{n}\,x^{-\frac{m}{n}-1}$.

The rule is therefore true for all powers, whether the exponent be positive or negative, integral or fractional.

(18.) The differential of any constant power of a function is found by multiplying by the exponent, diminishing the exponent by unity, and finally multiplying by the differential of the function.

Let $u = P^n$, P being a function of x; then proceeding as in article (17), only substituting P in place of x, we obtain

$$\frac{du}{dP} = n\,P^{n-1} \text{ and } du = n\,P^{n-1}\,dP.$$

As in the former case, this rule is also true for all powers, whether the exponent be positive or negative, integral or fractional.

Cor. Hence also $\dfrac{du}{dx} = n\,P^{n-1}\,\dfrac{dP}{dx}$

$$\text{and } du = n\,P^{n-1}\,\frac{dP}{dx}\,dx.$$

(19.) The differential of a function consisting of two variable factors is found by multiplying each factor by the differential of the other, and adding together the two products.

Let $u = P\,Q$, the factors P and Q being functions of x. When x becomes $x + \Delta x$ the corresponding values of u, P, Q will be $u + \Delta u$, $P + \Delta P$, $Q + \Delta Q$ respectively, and then

$$u + \Delta u = (P + \Delta P)(Q + \Delta Q) = P\,Q + Q\,\Delta P$$
$$+ (P + \Delta P)\,\Delta Q$$

$$\therefore \quad \Delta u = Q \, \Delta P + (P + \Delta P) \, \Delta Q$$

$$\frac{\Delta u}{\Delta x} = Q \, \frac{\Delta P}{\Delta x} + (P + \Delta P) \, \frac{\Delta Q}{\Delta x}.$$

Hence, making the increments vanish and taking the limiting values, we get

$$\frac{du}{dx} = Q \frac{dP}{dx} + P \frac{dQ}{dx} \text{ or } du = Q \, dP + P \, dQ.$$

(20.) The differential of a function consisting of any number of variable factors is found by adding together the products formed by multiplying the differential of each of the factors by all the others.

Let $u = P \, Q \, R$, a function consisting of three variable factors P, Q, R. By considering the function u to consist of two factors P Q and R, we have by (19)

$$du = R \, d(P \, Q) + P \, Q \, dR$$
$$= R \, (Q \, dP + P \, dQ) + P \, Q \, dR$$
$$= Q \, R \, dP + R \, P \, dQ + P \, Q \, dR.$$

Similarly if $u = P \, Q \, R \, S$, the product of four factors, we obtain

$$du = S \, d(P \, Q \, R) + P \, Q \, R \, dS$$
$$= S \, (Q \, R \, dP + R \, P \, dQ + P \, Q \, dR) + P \, Q \, R \, dS$$
$$= Q \, R \, S \, dP + R \, S \, P \, dQ + S \, P \, Q \, dR + P \, Q \, R \, dS;$$

and the same process of derivation may evidently be extended to any number of factors.

(21.) The differential of a function in the form of a fraction is found by multiplying the differential of the numerator by the denominator, from this product subtracting the differential of the denominator multiplied by the numerator, and dividing the remainder by the square of the denominator.

Let $u = \dfrac{P}{Q}$, P and Q being functions of x;

then $u + \Delta u = \dfrac{P + \Delta P}{Q + \Delta Q}$, and

$$\Delta u = \frac{P + \Delta P}{Q + \Delta Q} - \frac{P}{Q} = \frac{Q \Delta P - P \Delta Q}{Q(Q + \Delta Q)}$$

$$\therefore \frac{\Delta u}{\Delta x} = \frac{Q \dfrac{\Delta P}{\Delta x} - P \dfrac{\Delta Q}{\Delta x}}{Q(Q + \Delta Q)}.$$

Hence taking the limiting values when $\Delta x = 0$, we obtain

$$\frac{du}{dx} = \frac{Q \dfrac{dP}{dx} - P \dfrac{dQ}{dx}}{Q^2} \text{ or } du = \frac{Q\,dP - P\,dQ}{Q^2}.$$

The different forms of functions, considered in the foregoing articles (14) to (21), comprise all the combinations of quantity that can be effected by the ordinary operations of Algebra, and they will therefore enable us to differentiate all algebraical functions, however complicated. We shall now apply them to a few examples.

1. Let it be required to differentiate $u = 3x + 2a$.

Here, by (14) we must disregard the constant term $2a$, and by (15) we have $\dfrac{du}{dx} = 3$ or $du = 3\,dx$.

2. Differentiate $u = \dfrac{1}{x}$.

This being written $u = x^{-1}$, we have by (17),

$$\frac{du}{dx} = -1 \times x^{-1-1} = -x^{-2} = -\frac{1}{x^2}, \text{ or } du = -\frac{dx}{x^2}.$$

3. Differentiate $u = 2x^4 + ax^3 - 3a^2x^2$.
By (15) and (17),

$$\frac{du}{dx} = 2\frac{d(x^4)}{dx} + a\frac{d(x^3)}{dx} - 3a^2\frac{d(x^2)}{dx}$$

$$= 2(4x^3) + a(3x^2) - 3a^2(2x)$$

$$= 8x^3 + 3ax^2 - 6a^2x.$$

4. Differentiate $u = 4x^{\frac{3}{2}}$.

Here $\dfrac{du}{dx} = 4\,\dfrac{d\,(x^{\frac{3}{2}})}{dx} = 4\,(\tfrac{3}{2}\,x^{\frac{3}{2}-1}) = 6\,x^{\frac{1}{2}} = 6\,\sqrt{x}.$

5. Differentiate $u = (a + x)\,(b + x).$

By (14) and (19),

$$du = (b + x)\,dx + (a + x)\,dx = (a + b + 2\,x)\,dx$$

$$\text{or } \dfrac{du}{dx} = a + b + 2\,x.$$

6. Differentiate $u = (x - 2)^2\,(x^2 + 3).$

By (18) and (19),

$$du = (x^2 + 3) \times 2\,(x - 2)\,dx + (x - 2)^2 \times 2\,x\,dx$$
$$= 2\,(x - 2)\,(2\,x^2 - 2\,x + 3)\,dx.$$

$$\therefore\ \dfrac{du}{dx} = 2\,(x - 2)\,(2\,x^2 - 2\,x + 3).$$

7. Differentiate $u = a^m x^n + b^n x^m.$

By (15) and (17),

$$\dfrac{du}{dx} = a^m\,(n\,x^{n-1}) + b^n\,(m\,x^{m-1}) = n\,a^m x^{n-1} + m\,b^n x^{m-1}.$$

8. Differentiate $u = (a + x)\,(b + 2x)\,(c + 3x).$

By (20) we have

$$du = (b + 2x)\,(c + 3x)\,.\,dx + (c + 3x)\,(a + x)\,.\,2\,dx$$
$$+ (a + x)\,(b + 2x)\,.\,3\,dx$$

$$\therefore\ \dfrac{du}{dx} = (b + 2x)\,(c + 3x) + 2\,(c + 3x)\,(a + x)$$
$$+ 3\,(a + x)\,(b + 2x)$$
$$= (3\,ab + bc + 2\,ca) + (12\,a + 6\,b + 4\,c)\,x + 18\,x^2.$$

9. Differentiate $u = \dfrac{a + x}{a - x}.$

By (21),

$$du = \dfrac{(a - x) \times dx - (a + x) \times - dx}{(a - x)^2}$$

$$= \frac{(a - x)\, dx + (a + x)\, dx}{(a - x)^2} = \frac{2\, a\, dx}{(a - x)^2}$$

$$\therefore \frac{du}{dx} = \frac{2\, a}{(a - x)^2}.$$

10. Differentiate $u = \sqrt{\dfrac{a + x}{a - x}}$, or $u = \dfrac{(a + x)^{\frac{1}{2}}}{(a - x)^{\frac{1}{2}}}$.

Here $du =$

$$\frac{(a - x)^{\frac{1}{2}} \times \frac{1}{2}(a + x)^{-\frac{1}{2}}\, dx - (a + x)^{\frac{1}{2}} \times -\frac{1}{2}(a - x)^{-\frac{1}{2}}\, dx}{(a - x)}$$

$$\therefore \frac{du}{dx} = \frac{(a - x)^{\frac{1}{2}}(a + x)^{-\frac{1}{2}} + (a + x)^{\frac{1}{2}}(a - x)^{-\frac{1}{2}}}{2(a - x)}$$

$$= \frac{(a - x) + (a + x)}{2(a - x)(a - x)^{\frac{1}{2}}(a + x)^{\frac{1}{2}}} = \frac{a}{(a - x)^{\frac{3}{2}}(a + x)^{\frac{1}{2}}}$$

$$= \frac{a}{(a - x)\sqrt{a^2 - x^2}}.$$

Otherwise, by squaring, we have $u^2 = \dfrac{a + x}{a - x}$ and, by the

last example, $2u\, du = \dfrac{2\, a\, dx}{(a - x)^2}$;

$$\therefore \frac{du}{dx} = \frac{a}{u(a - x)^2} = \frac{a}{(a - x)^2}\sqrt{\frac{a - x}{a + x}}$$

$$= \frac{a}{(a - x)\sqrt{a^2 - x^2}}.$$

11. Differentiate $u = \sqrt{a^2 - x^2}$.

Write $u = (a^2 - x^2)^{\frac{1}{2}}$ and by (17), (18),

$$du = \frac{1}{2}(a^2 - x^2)^{-\frac{1}{2}} \times -2x\, dx = \frac{-x\, dx}{\sqrt{a^2 - x^2}}.$$

12. Differentiate $u = \sqrt{a^2 + 2bx + x^2}$.

Here $u = (a^2 + 2bx + x^2)^{\frac{1}{2}}$;

$$\therefore du = \frac{1}{2}(a^2 + 2bx + x^2)^{-\frac{1}{2}} \times (2b\, dx + 2x\, dx)$$

$$= \frac{(b + x)\, dx}{\sqrt{a^2 + 2bx + x^2}}.$$

13. Differentiate $u = \dfrac{\sqrt{(a^2 + x^2)^3}}{x^3} = \dfrac{(a^2 + x^2)^{\frac{3}{2}}}{x^3}$.

By (18) and (21)

$$du = \frac{x^3 \times 3(a^2 + x^2)^{\frac{1}{2}} x\, dx - (a^2 + x^2)^{\frac{3}{2}} \times 3\, x^2\, dx}{x^6}$$

$$\therefore \frac{du}{dx} = \frac{3\, x^2\, (a^2 + x^2)^{\frac{1}{2}} \left\{ x^2 - (a^2 + x^2) \right\}}{x^6}$$

$$= - \frac{3\, a^2}{x^4} \sqrt{a^2 + x^2}.$$

Otherwise, writing the function in the form

$u = (a^2 + x^2)^{\frac{3}{2}} x^{-3}$, we obtain by (19)

$$du = x^{-3} \times 3\, x\, dx\, (a^2 + x^2)^{\frac{1}{2}} + (a^2 + x^2)^{\frac{3}{2}} \times -3\, x^{-4}$$

$$= 3\, dx\, (a^2 + x^2)^{\frac{1}{2}} \left\{ x^{-2} - x^{-4} (a^2 + x^2) \right\}$$

$$= - 3\, a^2\, x^{-4}\, dx\, (a^2 + x^2)^{\frac{1}{2}} = - \frac{3\, a^2}{x^4} \sqrt{a^2 + x^2}.$$

14. Differentiate $u = \dfrac{x}{\sqrt{a^2 - x^2}} = \dfrac{x}{(a^2 - x^2)^{\frac{1}{2}}}$.

$$du = \frac{(a^2 - x^2)^{\frac{1}{2}} \times dx - x \times -(a^2 - x^2)^{-\frac{1}{2}} x\, dx}{a^2 - x^2}$$

$$= \frac{(a^2 - x^2)\, dx + x^2\, dx}{(a^2 - x^2)^{\frac{3}{2}}} = \frac{a^2\, dx}{(a^2 - x^2)^{\frac{3}{2}}}.$$

15. Differentiate $u = \dfrac{\sqrt{a + x} - \sqrt{a - x}}{\sqrt{a + x} + \sqrt{a - x}}$.

Differential of the numerator

$$= \tfrac{1}{2}(a + x)^{-\frac{1}{2}}\, dx + \tfrac{1}{2}(a - x)^{-\frac{1}{2}}\, dx$$

$$= \frac{\sqrt{a + x} + \sqrt{a - x}}{2\sqrt{a^2 - x^2}}\, dx.$$

Differential of the denominator

$$= \tfrac{1}{2}(a+x)^{-\frac{1}{2}}dx - \tfrac{1}{2}(a-x)^{-\frac{1}{2}}dx$$

$$= -\frac{\sqrt{a+x} - \sqrt{a-x}}{2\sqrt{a^2-x^2}}dx.$$

Therefore by (21) we have

$$du = \frac{(\sqrt{a+x} + \sqrt{a-x})^2 + (\sqrt{a+x} - \sqrt{a-x})^2}{2(\sqrt{a+x} + \sqrt{a-x})^2 \sqrt{a^2-x^2}}dx$$

$$= \frac{a\,dx}{(a+\sqrt{a^2-x^2})\sqrt{a^2-x^2}} = \frac{a(a-\sqrt{a^2-x^2})}{x^2\sqrt{a^2-x^2}}dx.$$

16. Differentiate $u = \dfrac{8a^4 - 4a^2x^2 - x^4}{\sqrt{a^2-x^2}}.$

Writing $(a^2-x^2)^{\frac{1}{2}}$ for $\sqrt{a^2-x^2}$, we similarly have, by (21),

$$du =$$

$$\frac{(a^2-x^2)^{\frac{1}{2}} \times (-8a^2x-4x^3)\,dx - (8a^4-4a^2x^2-x^4) \times -(a^2-x^2)^{-\frac{1}{2}}x\,dx}{a^2-x^2}$$

$$= \frac{-(a^2-x^2)(8a^2x+4x^3) + (8a^4-4a^2x^2-x^4)x}{(a^2-x^2)^{\frac{3}{2}}}dx = \frac{3x^5\,dx}{(a^2-x^2)^{\frac{3}{2}}}.$$

17. If $u = (a-x)(b+x)$; then $\dfrac{du}{dx} = a - b$.

18. If $u = \dfrac{1}{x} + \dfrac{3}{x^2} + \dfrac{3}{x^3}$; then $\dfrac{du}{dx} = -\left(\dfrac{x+3}{x^2}\right)^2.$

19. If $u = (a^2 + bx + x^2)\sqrt{x}$; then $\dfrac{du}{dx} = \dfrac{a^2 + 3bx + 5x^2}{2\sqrt{x}}.$

20. If $u = (2 + x^2)\sqrt{1-x^2}$; then $\dfrac{du}{dx} = -\dfrac{3x^3}{\sqrt{1-x^2}}.$

21. If $u = \dfrac{2x^2 - a^2}{x^3}\sqrt{a^2+x^2}$; then $\dfrac{du}{dx} = \dfrac{3a^4}{x^4\sqrt{a^2+x^2}}.$

22. If $u = \dfrac{(a^2 + x^2)^{\frac{3}{2}}}{x^3}$; then $\dfrac{du}{dx} = -\dfrac{3\,a^2}{x^4}\sqrt{a^2 + x^2}$.

23. If $u = (3\,x^2 - 2\,a^2)\,(a^2 + x^2)^{\frac{3}{2}}$;

$$\text{then } \frac{du}{dx} = 15\,x^3\sqrt{a^2 + x^2}.$$

(22.) Expressions under the form of square roots are of very frequent occurrence in analytical investigations, and their differentiation, according to art. (18), using $\frac{1}{2}$ for the exponent, suggests the following simple and expeditious rule :

The differential of the square root of a function is found by taking half the differential of the function and dividing the same by the square root of the function.

This useful rule may be practically applied by the student to Nos. 11, 12, 14, 16, 20, 21, of the preceding examples, and it will enable him at once to put down the final result in all ordinary cases of this kind.

II. *Logarithmic and Exponential Functions.*

(23.) The logarithmic function $u = \log x$ depends upon the exponential relation $a^u = a^{\log x} = x$. Thus if $a^{\log x} = x$, and $a^{\log y} = y$, we have, by multiplication, $a^{\log x + \log y} = x\,y$; but $a^{\log(xy)} = x\,y$,

$$\therefore \ \log x + \log y = \log(x\,y),$$

which is the fundamental property of logarithms.

The constant quantity a is indeterminate and may have any proposed value. It is called the base of the logarithmic system belonging to it, and, since $a^1 = a$, it is evidently the number whose logarithm in the same system is equal to unity.

Since $x = a^u$, we have $x + \Delta x = a^{u + \Delta u}$, and therefore

$$\frac{\Delta x}{\Delta u} = \frac{a^{u + \Delta u} - a^u}{\Delta u} = a^u \cdot \frac{a^{\Delta u} - 1}{\Delta u}.$$

In taking the limits of this equation we observe that the

limiting continuous value of the fraction $\dfrac{a^{\Delta u} - 1}{\Delta u}$, which in

common with $\dfrac{\Delta x}{\Delta u}$ takes the form $\dfrac{0}{0}$ when $\Delta u = 0$, must be a

function of a and independent of Δu. Denoting this function by $\lambda\, a$, we have

$$\lambda\, a = \text{limiting value of } \frac{a^\theta - 1}{\theta} \text{ when } \theta = 0$$

$$\frac{dx}{du} = a^u \lambda\, a = x \lambda\, a.$$

Again, the equation $x = a^u$ gives $x^\theta = a^{u\theta}$, θ denoting any value whatever. Therefore

$$\frac{x^\theta - 1}{\theta} = \frac{a^{u\theta} - 1}{\theta} = u \cdot \frac{a^{u\theta} - 1}{u\,\theta}.$$

This equation is necessarily true for all values of θ. By proceeding to the limit $\theta = 0$, $u\,\theta = 0$, the continuous values, from what precedes, obviously give

$$\lambda\, x = u \lambda\, a ;$$

$$\therefore\ u = \log x = \frac{\lambda\, x}{\lambda\, a}.$$

The value of the function $\lambda\, x$ may readily be obtained in a

series by putting $\dfrac{x^\theta - 1}{\theta}$ in the form $\dfrac{\{1 + (x-1)\}^\theta - 1}{\theta}$.

Thus, by expanding according to the binomial theorem and putting $\theta = 0$ in the final result, we obtain

$$\lambda\, x = (x - 1) - \tfrac{1}{2}(x-1)^2 + \tfrac{1}{3}(x-1)^3 - \tfrac{1}{4}(x-1)^4 + \&c.,$$

so that the last expression for $\log x$ may be written

$$\log x = \frac{(x-1) - \tfrac{1}{2}(x-1)^2 + \tfrac{1}{3}(x-1)^3 - \tfrac{1}{4}(x-1)^4 + \&c.}{(a-1) - \tfrac{1}{2}(a-1)^2 + \tfrac{1}{3}(a-1)^3 - \tfrac{1}{4}(a-1)^4 + \&c.}.$$

These equations apply generally to a system of logarithms having any value a for the base. According to Briggs's system, on which the logarithmic tables in common use have been calculated, the base $a = 10$, which greatly facilitates the use of the tables in arithmetical calculations which involve decimal numbers.

(24.) If the value of a be so assigned that $\lambda a = 1$, we shall have $\log x = \lambda x$, and $\log a = \lambda a = 1$. This value of a will simplify the analytical relations and give the Napierian system of logarithms, of which the, value of a so determined is the base. Hence it follows that the function we have indicated by λ characterizes the Napierian logarithm. To determine the particular value of a which will fulfil the proposed condition $\lambda a = 1$, instead of using the series for λa take the initial form of this function, and we have

$$\text{limit of } \frac{a^\theta - 1}{\theta} = 1, \text{ when } \theta = 0 ;$$

$$\therefore \ a = \text{limit of } (1 + \theta)^{\frac{1}{\theta}}, \text{ when } \theta = 0.$$

By expanding according to the binomial theorem, we find

$$(1 + \theta)^{\frac{1}{\theta}} = 1 + \frac{1}{\theta}\theta + \frac{\frac{1}{\theta}\left(\frac{1}{\theta} - 1\right)}{2}\theta^2$$

$$+ \frac{\frac{1}{\theta}\left(\frac{1}{\theta} - 1\right)\left(\frac{1}{\theta} - 2\right)}{2.3}\theta^3 + \&c.$$

$$= 1 + 1 + \frac{1 - \theta}{2} + \frac{(1 - \theta)(1 - 2\theta)}{2.3} + \&c.$$

Now, when θ passes from a small positive to a small negative value, the value of every term of this series will evidently vary continuously, and when $\theta = 0$ it gives the limiting value of $(1 + \theta)^{\frac{1}{\theta}}$

$$= 1 + 1 + \frac{1}{2} + \frac{1}{2.3} + \frac{1}{2.3.4} + \&c. = 2\cdot7182818, \&c.$$

This arithmetical value, which forms the base of the Napierian logarithms, is usually denoted by the letter e, and sometimes by the Greek letter ϵ, and these symbols *always* represent this arithmetical value whenever they appear as roots of exponential functions.

The Napierian system, from its greater algebraical simplicity and convenience, is also that which is generally employed in analytical investigations and formulæ ; and therefore whenever a logarithmic expression occurs, the Napierian logarithm should always be understood unless the contrary is distinctly stated. We have thus, according to this system, the following relations :

$$\log x = \text{limit of } \frac{x^\theta - 1}{\theta}, \text{ when } \theta = 0.$$

$$e = \text{limit of } (1 + \theta)^{\frac{1}{\theta}} = 2\cdot7182818, \&c.$$

$$e^{\log x} = x.$$

When $u = \log x$, the expression for $\dfrac{dx}{du}$ (art. 23) also becomes $\dfrac{dx}{du} = x$, giving $du = \dfrac{dx}{x}$; but we shall otherwise determine this differentiation in the next article.

(25.) Differentiation of $u = \log x$.

When x becomes $x + \Delta x$, u becomes $u + \Delta u$, and we have

$$u + \Delta u = \log (x + \Delta x) ;$$

$$\therefore \ \Delta u = \log (x + \Delta x) - \log x = \log \frac{x + \Delta x}{x} = \log \left(1 + \frac{\Delta x}{x} \right)$$

and, putting $\dfrac{\Delta x}{x} = \theta$, we find

$$\frac{\Delta u}{\Delta x} = \frac{\log (1 + \theta)}{x \theta} = \frac{1}{x} \log \left\{ (1 + \theta)^{\frac{1}{\theta}} \right\}$$

In proceeding to the limit $\Delta u = 0$, $\Delta x = 0$, $\theta = 0$, we observe that the continuous limiting value of $(1 + \theta)^{\frac{1}{\theta}} = e$ and that $\log e = 1$. Hence

$$\frac{du}{dx} = \frac{1}{x}, \text{ and } du = \frac{dx}{x}.$$

Therefore the differential of the logarithm of a variable quantity is found by taking the differential of the quantity and dividing by the quantity itself.

The differential of a power, or of the product of several functions, may be readily deduced from this. Thus if $u = x^n$, then $\log u = n \log x$, the differential of which gives $\dfrac{du}{u} = n \dfrac{dx}{x}$;

$\therefore \dfrac{du}{dx} = n \dfrac{u}{x} = n x^{n-1}$, the same as in art. (17). Again, if $u = P \times Q \times R$, &c., then $\log u = \log P + \log Q + \log R +$ &c., and $\therefore \dfrac{du}{u} = \dfrac{dP}{P} + \dfrac{dQ}{Q} + \dfrac{dR}{R} +$ &c., which gives

$$du = u \left(\frac{dP}{P} + \frac{dQ}{Q} + \frac{dR}{R} + \&c \right.$$

$$= PQR, \&c. \left(\frac{dP}{P} + \frac{dQ}{Q} + \frac{dR}{R} + \&c. \right)$$

which is equivalent to the formula of art. (20).

(26.) Differentiation of $u = a^x$.

When x becomes $x + \Delta x$ we have $u + \Delta u = a^{x + \Delta x}$;

$$\therefore \frac{\Delta u}{\Delta x} = \frac{a^{x + \Delta x} - a^x}{\Delta x} = a^x \frac{a^{\Delta x} - 1}{\Delta x}.$$

But (art. 24) the limiting value of the vanishing fraction $\dfrac{a^{\Delta x} - 1}{\Delta x}$, which is of the form $\dfrac{a^\theta - 1}{\theta}$, is $\log a$; therefore

$$\frac{du}{dx} = \log a \cdot a^x, \quad \text{or } du = \log a \cdot a^x \, dx.$$

Thus the differential of an exponential quantity having an invariable root is found by multiplying together the logarithm of the root, the exponential itself, and the differential of its exponent.

Hence, when $a = e$, or $u = e^x$, we have, since $\log e = 1$,

$$\frac{du}{dx} = e^x, \quad \text{or } du = e^x \, dx;$$

that is, the differential of an exponential quantity having for its root the Napierian base e is found by multiplying it by the differential of the exponent.

(27.) Differentiation of $u = P^Q$, P and Q being functions of x.

Since $u = P^Q$, we have $\log u = Q \log P$, the differential of which gives $d(\log u) = (\log P) dQ + Qd(\log P)$; that is, by (25),

$$\frac{du}{u} = (\log P) dQ + Q \frac{dP}{P};$$

$$\therefore \ du = (\log P) u \, dQ + Q u \frac{dP}{P}$$

$$= (\log P) P^Q dQ + Q P^{Q-1} dP.$$

Hence the differential of an exponential quantity when the root and exponent are both variable is found by adding together the differentials obtained by considering each separately as constant and the other variable.

For example, let $u = x^{x^2}$. By considering the root x to be constant and the exponent x^2 to be variable, we obtain by (26) the differential $(\log x) x^{x^2} \times 2x \, dx = 2 x^{x^2+1} dx (\log x)$. Again, by considering the exponent x^2 to be constant and the root x to be variable, we obtain by (17) the differential $x^2 . x^{x^2-1} dx = x^{x^2+1} dx$. Hence, adding these, we find

$$du = x^{x^2+1} dx \,(2 \log x + 1) \quad \text{or} \quad \frac{du}{dx} = x^{x^2+1} \,(2 \log x + 1).$$

The following examples are added as exercises :

1. If $u = x^m e^x$; then $\dfrac{du}{dx} = x^{m-1} (m + x) e^x$.

2. If $u = (x^2 - 2x + 2) e^x$; then $\dfrac{du}{dx} = x^2 e^x$.

3. If $u = (x^3 - 3x^2 + 6x - 6) e^x$; then $\dfrac{du}{dx} = x^3 e^x$.

4. If $u = \dfrac{e^x}{1 + x}$; then $\dfrac{du}{dx} = \dfrac{x e^x}{(1 + x)^2}$.

5. If $u = e^x \log x$; then $\dfrac{du}{dx} = e^x \left(\dfrac{1}{x} + \log x \right)$.

6. If $u = e^{mx} \log x$; then $\dfrac{du}{dx} = e^{mx} \left(\dfrac{1}{x} + m \log x \right)$.

7. If $u = e^x \sqrt{1 + x^2}$; then $\dfrac{du}{dx} = \dfrac{1 + x + x^2}{\sqrt{1 + x^2}} e^x$.

III. *Trigonometrical Functions.*

(28.) The trigonometrical functions sin x, cos x, tan x, &c. are usually considered as abstract arithmetical quantities having reference to a circle whose radius is *unity ;* or, which is in effect the same, they are supposed to be expressed in parts of the radius, the arithmetical value of the variable x being supposed to represent the length of the arc measured on a circle whose radius is unity or otherwise expressed in parts of the radius of the circle. Other forms result from the various combinations of these elementary functions, and as they all involve relations between arcs of circles and their coordinates they are sometimes called " circular functions."

1. Differentiation of $u = \sin x$.

When x becomes $x + \Delta x$, then $u + \Delta u = \sin (x + \Delta x)$, and

$$\Delta u = \sin (x + \Delta x) - \sin x$$

$$= \sin \{ (x + \tfrac{1}{2} \Delta x) + \tfrac{1}{2} \Delta x \}$$

$$- \sin \{ (x + \tfrac{1}{2} \Delta x) - \tfrac{1}{2} \Delta x \}$$

$$= 2 \cos (x + \tfrac{1}{2} \Delta x) \sin \tfrac{1}{2} \Delta x$$

$$= \cos (x + \tfrac{1}{2} \Delta x) \operatorname{ch} \Delta x ;$$

$$\therefore \ \frac{\Delta u}{\Delta x} = \cos (x + \tfrac{1}{2} \Delta x) \frac{\operatorname{ch} \Delta x}{\Delta x}.$$

Now, when Δu and Δx become infinitesimals, or when we suppose $\Delta x = 0$ with the view of seeking the limit of this equation, the fraction $\dfrac{\operatorname{ch} \Delta x}{\Delta x}$ becomes a vanishing fraction, and therefore it will first be requisite to ascertain its limiting value. Let ch Δx be considered to be the side of a regular polygon of n sides inscribed within the circle, and we shall obviously have

$$\frac{\text{ch } \Delta x}{\Delta x} = \frac{n \text{ ch } \Delta x}{n \Delta x} = \frac{\text{perimeter of polygon}}{\text{periphery of circle}}.$$

If the number of sides of the polygon be supposed to be indefinitely increased, so will Δx become indefinitely diminished, and the perimeter of the polygon will evidently approximate more and more nearly to the circumference of the circle as its extreme limit, so that the numerator of the fraction $\frac{\text{perimeter of polygon}}{\text{periphery of circle}}$ will ultimately become equal to the denominator; and thus the limiting value of $\frac{\text{ch } \Delta x}{\Delta x}$ is $\frac{\text{ch } dx}{dx}$ = *unity*. Therefore by supposing $\Delta x = 0$ and taking the limit of the preceding value of $\frac{\Delta u}{\Delta x}$ we obtain the ultimate differential relation

$$\frac{du}{dx} = \cos x, \quad \text{or } du = dx \cos x.$$

Cor. The limiting value of $\frac{\sin \theta}{\theta} = 1$, when θ vanishes.

For $\frac{\sin \theta}{\theta} = \frac{\frac{1}{2} \text{ch } 2\theta}{\theta} = \frac{\text{ch } 2\theta}{2\theta}$, which is of the same form as $\frac{\text{ch } \Delta x}{\Delta x}$, and therefore expresses the same ratio in the limit.

2. Differentiation of $u = \cos x$.

Here $\Delta u = \cos (x + \Delta x) - \cos x$

$$= \cos \left\{ (x + \tfrac{1}{2} \Delta x) + \tfrac{1}{2} \Delta x \right\}$$
$$- \cos \left\{ (x + \tfrac{1}{2} \Delta x) - \tfrac{1}{2} \Delta x \right\}$$
$$= - 2 \sin (x + \tfrac{1}{2} \Delta x) \sin \tfrac{1}{2} \Delta x$$
$$= - \sin (x + \tfrac{1}{2} \Delta x) \text{ ch } \Delta x ;$$

$$\therefore \frac{\Delta u}{\Delta x} = - \sin (x + \tfrac{1}{2} \Delta x) \frac{\text{ch } \Delta x}{\Delta x}.$$

Hence, taking the limit as before,

$$\frac{du}{dx} = -\sin x, \quad \text{or } du = -dx \sin x.$$

Otherwise, since $u = \cos x = \sin\left(\frac{1}{2}\pi - x\right)$, we have

$$du = d\left(\tfrac{1}{2}\pi - x\right)\cos\left(\tfrac{1}{2}\pi - x\right)$$
$$= -dx \cos\left(\tfrac{1}{2}\pi - x\right) = -dx \sin x.$$

3. Differentiation of $u = \tan x$.

Since $u = \tan x = \dfrac{\sin x}{\cos x}$, we have, by (21),

$$du = \frac{\cos x \, d\sin x - \sin x \, d\cos x}{\cos^2 x}$$

$$= \frac{\cos x \, (dx \cos x) - \sin x \, (-dx \sin x)}{\cos^2 x}$$

$$= \frac{dx \, (\cos^2 x + \sin^2 x)}{\cos^2 x} = \frac{dx}{\cos^2 x} = dx \sec^2 x.$$

4. Differentiation of $u = \cot x$.

Here $u = \cot x = \dfrac{\cos x}{\sin x}$, and

$$du = \frac{\sin x \, d\cos x - \cos x \, d\sin x}{\sin^2 x}$$

$$= \frac{\sin x \, (-dx \sin x) - \cos x \, (dx \cos x)}{\sin^2 x}$$

$$= -\frac{dx \, (\sin^2 x + \cos^2 x)}{\sin^2 x}$$

$$= -\frac{dx}{\sin^2 x} = -dx \csc^2 x.$$

Otherwise, since $u = \cot x = \dfrac{1}{\tan x}$, we have, according to example 2, page 22, and the preceding,

$$du = -\frac{d\tan x}{\tan^2 x} = -\frac{dx \sec^2 x}{\tan^2 x} = -\frac{dx}{\sin^2 x} = -dx \csc^2 x.$$

Or this differentiation may be obtained from that of tan x by putting $u = \cot x = \tan(\frac{1}{2}\pi - x)$; thus we have

$$du = d(\tfrac{1}{2}\pi - x)\sec^2(\tfrac{1}{2}\pi - x) = -dx\sec^2(\tfrac{1}{2}\pi - x)$$
$$= -dx\operatorname{cosec}^2 x.$$

5. Differentiation of $u = \sec x$.

Since $u = \sec x = \dfrac{-}{\cos x}$, we have

$$du = -\frac{d\cos x}{\cos^2 x} = \frac{dx\sin x}{\cos^2 x} = dx\tan x\sec x.$$

6. Differentiation of $u = \operatorname{cosec} x$.

Here $u = \operatorname{cosec} x = \dfrac{1}{\sin x}$, and

$$du = -\frac{d\sin x}{\sin^2 x} = -\frac{dx\cos x}{\sin^2 x} = -dx\cot x\operatorname{cosec} x.$$

Otherwise, since $u = \operatorname{cosec} x = \sec(\frac{1}{2}\pi - x)$, we have, by the preceding,
$$du = d(\tfrac{1}{2}\pi - x)\tan(\tfrac{1}{2}\pi - x)\sec(\tfrac{1}{2}\pi - x) = -dx\cot x\operatorname{cosec} x.$$

(29.) The differentiation of other more complicated trigonometrical functions may be easily deduced from the elementary differentials here obtained, because all such functions must evidently result from certain combinations of these with algebraic functions. As it may therefore be useful to remember the results of the preceding trigonometrical differentiations, it will be convenient to collect them together as follows:

$d\sin x = dx\cos x$	$d\cos x = -dx\sin x$
$d\tan x = dx\sec^2 x$	$d\cot x = -dx\operatorname{cosec}^2 x$
$d\sec x = dx\tan x\sec x.$	$d\operatorname{cosec} x = -dx\cot x\operatorname{cosec} x.$

They are thus arranged in two columns because the differentials in the second column are respectively analogous to those in the first column, only using the complementary angle or substituting $\frac{1}{2}\pi - x$ in place of x; and, this analogy being once recognized, a remembrance of the three differentials in the first column will be sufficient to suggest the others.

Examples for exercise :

1. If $u = \cos x + x \sin x$; then $\dfrac{du}{dx} = x \cos x$.

2. If $u = \cos^m x \sin^n x$;

then $\dfrac{du}{dx} = \cos^{m-1} x \sin^{n-1} x \, (n \cos^2 x - m \sin^2 x)$.

3. If $u = (2 + \cos^2 x) \sin x$; then $\dfrac{du}{dx} = 3 \cos^3 x$.

4. If $u = 2 x \sin x + (2 - x^2) \cos x$; then $\dfrac{du}{dx} = x^2 \sin x$.

5. If $u = (2 + 3 \cos^2 x) \sin^3 x$; then $\dfrac{du}{dx} = 15 \cos^3 x \sin^2 x$.

6. If $u = 3 x - 3 \tan x + \tan^3 x$; then $\dfrac{du}{dx} = 3 \tan^4 x$.

7. If $u = 2 \cos x + 2 x \sin x - x' \cos x$; then $\dfrac{du}{dx} = x^2 \sin x$.

8. If $u = 3 x - \cos x \, (3 \sin x + 2 \sin^3 x)$; then $\dfrac{du}{dx} = 8 \sin^4 x$.

9. If $u = e^x (\cos x + \sin x)$; then $\dfrac{du}{dx} = 2 e^x \cos x$.

IV. *Inverse Functions.*

(30.) If $x = fu$, a function of u, the reverse relation which indicates the corresponding value of u as depending upon that of x is called an *inverse function*, and is usually written $u = f^{-1}x$. Thus if $x = \sin u$, then $u = \sin^{-1}x$, and this inverse trigonometrical function therefore symbolically expresses the circular arc whose sine is x. Similarly $u = \log^{-1}x$ expresses the number whose Napierian logarithm is equal to x. The differentiation of an inverse function follows immediately from that of the direct function. For, taking $u = f^{-1}x$, we have $x = fu$, the differential of which gives $dx = du f'u$,

$$\therefore \quad \frac{du}{dx} = \frac{1}{f'u} = \frac{1}{f'(f^{-1}x)}.$$

We shall here in this way determine the differentials of the ordinary inverse functions in their simplest form.

1. Differentiation of $u = \log^{-1} x$.

Since $x = \log u$, we have by (25) $dx = \dfrac{du}{u}$;

$$\therefore \frac{du}{dx} = u = \log^{-1} x, \quad \text{or } du = dx \log^{-1} x.$$

2. Differentiation of $u = \sin^{-1} x$.

Since $x = \sin u$, we have by (29) $dx = du \cdot \cos u$;

$$\therefore \frac{du}{dx} = \frac{1}{\cos u} = \frac{1}{\sqrt{1 - x^2}}, \quad \text{or } du = \frac{dx}{\sqrt{1 - x^2}}.$$

3. Differentiation of $u = \cos^{-1} x$.

Since $x = \cos u$, we have $dx = - du \sin u$;

$$\therefore \frac{du}{dx} = - \frac{1}{\sin u} = - \frac{1}{\sqrt{1 - x^2}}, \quad \text{or } du = - \frac{dx}{\sqrt{1 - x^2}}.$$

4. Differentiation of $u = \tan^{-1} x$.

Since $x = \tan u$, we have $dx = du \sec^2 u$;

$$\therefore \frac{du}{dx} = \frac{1}{\sec^2 u} = \frac{1}{1 + x^2}, \quad \text{or } du = \frac{dx}{1 + x^2}.$$

5. Differentiation of $u = \cot^{-1} x$.

Since $x = \cot u$, we have $dx = - du \csc^2 u$;

$$\therefore \frac{du}{dx} = - \frac{1}{\csc^2 u} = - \frac{1}{1 + x^2}, \quad \text{or } du = - \frac{dx}{1 + x^2}.$$

6. Differentiation of $u = \sec^{-1} x$.

Since $x = \sec u$, we have $dx = du \tan u \sec u$;

$$\therefore \frac{du}{dx} = \frac{1}{\tan u \sec u} = \frac{1}{x \sqrt{x^2 - 1}}, \quad \text{or } du = \frac{dx}{x \sqrt{x^2 - 1}}.$$

7. Differentiation of $u = \csc^{-1} x$.

Since $x = \csc u$, we have $dx = - du \cot u \csc u$;

$$\therefore \frac{du}{dx} = - \frac{1}{\cot u \csc u} = - \frac{1}{x \sqrt{x^2 - 1}}$$

$$\text{or } du = - \frac{dx}{x \sqrt{x^2 - 1}}.$$

Here the differentials of $\cos^{-1}x$, $\cot^{-1}x$, $\operatorname{cosec}^{-1}x$ are respectively the same as the differentials of $\sin^{-1}x$, $\tan^{-1}x$, $\sec^{-1}x$ only with the negative sign; and this should evidently be the case, because $\frac{1}{2}\pi = \cos^{-1}x + \sin^{-1}x = \cot^{-1}x + \tan^{-1}x = \operatorname{cosec}^{-1}x + \sec^{-1}x$.

Examples for exercise :

1. If $u = (x^2 - 2x + 2)\log^{-1}x$; then $\dfrac{du}{dx} = x^2\log^{-1}x$.

2. If $u = \dfrac{\log^{-1}x}{1 + x}$; then $\dfrac{du}{dx} = \dfrac{x\log^{-1}x}{(1 + x)^2}$.

3. If $u = \log x \log^{-1}x$; then $\dfrac{du}{dx} = \left(\log x + \dfrac{1}{x}\right)\log^{-1}x$.

4. If $u = \tan^{-1}x + \dfrac{1}{x}$; then $\dfrac{du}{dx} = -\dfrac{1}{x^2(1 + x^2)}$.

5. If $u = \tan^{-1}x \sqrt{1 + x^2}$; then $\dfrac{du}{dx} = \dfrac{1 + x\tan^{-1}x}{\sqrt{1 + x^2}}$.

6. If $u = x - \sqrt{1 - x^2}\sin^{-1}x$; then $\dfrac{du}{dx} = \dfrac{x\sin^{-1}x}{\sqrt{1 - x^2}}$.

7. If $u = (2x^2 - 1)\sin^{-1}x + x\sqrt{1 - x^2}$;

then $\dfrac{du}{dx} = 4x\sin^{-1}x$.

8. If $u = x^2 + (\sin^{-1}x)^2 - 2\sin^{-1}x . x\sqrt{1 - x^2}$;

then $\dfrac{du}{dx} = \dfrac{4x^2\sin^{-1}x}{\sqrt{1 - x^2}}$.

v. *Compound Functions.*

(31.) If in a function $u = fx$ the variable x is replaced by another function ϕx, the expression $u = f(\phi x)$, which then becomes a function of a function, is called *a compound function* of x.

Let $y = \phi x$, so that $u = fy$, and let Δu, Δx, Δy denote corresponding increments of u, x, y; then, as the equation

$$\frac{\Delta u}{\Delta x} = \frac{\Delta u}{\Delta y} \cdot \frac{\Delta y}{\Delta x}$$

essentially represents an identity, and is therefore true for all values of the increments, however small, it must evidently be true when we proceed to the limit or suppose the increments to vanish and take the continuous values of the respective fractions. Hence

$$\frac{du}{dx} = \frac{du}{dy} \cdot \frac{dy}{dx}, \quad \text{or } du = \frac{du}{dy} \cdot \frac{dy}{dx} \cdot dx,$$

where $\frac{du}{dy}$, $\frac{dy}{dx}$ are the differential coefficients of the functions $u = fy$, and $y = \phi x$. That is, according to the usual notation of derived functions,

$$\frac{du}{dx} = f'y \cdot \phi'x = f'(\phi x)\, \phi'x,$$

$$\text{or } du = f'(\phi x)\, \phi'x.dx.$$

Similarly, if $y = \phi x$, $z = \psi y$, $u = fz$, so that the function u is of the more complicated form $u = f\{\psi(\phi x)\}$, or the function of a function of a function, it may be shown that

$$\frac{du}{dx} = \frac{du}{dz} \cdot \frac{dz}{dy} \cdot \frac{dy}{dx}, \quad \text{or } du = \frac{du}{dz} \cdot \frac{dz}{dy} \cdot \frac{dy}{dx} \cdot dx;$$

and these, according to the notation of derived functions, would be written

$$\frac{du}{dx} = f'z \cdot \psi'y \cdot \phi'x = f'(\psi y)\, \psi'(\phi x)\, \phi'x$$

$$= f'\{\psi(\phi x)\}\, \psi'(\phi x)\, \phi'x$$

$$\text{or } du = f'\{\psi(\phi x)\}\, \psi'(\phi x)\, \phi'x.dx.$$

In the same way the formulæ may be extended to any number of superposed functions, and it is obvious that they all depend upon the following simple principle:

The differential of a function of any variable quantity whatever is equal to the differential coefficient of the function, with respect to that variable quantity, multiplied by the differential of the variable quantity.

Thus if, as before, $u = f\{\psi(\phi x)\}$, by successively applying this principle, we have

$$du = f'\{\psi(\phi x)\} \times d\{\psi(\phi x)\}$$

$$= f'\{\psi(\phi x)\} \times \psi'(\phi x) \times d(\phi x)$$

$$= f'\{\psi(\phi x)\} \times \psi'(\phi x) \times \phi' x \times dx.$$

The following examples will practically show the mode of proceeding here indicated :

1. Differentiate $u = \log \sin x.$

By (25) and (29) we have

$$du = \frac{d \sin x}{\sin x} = \frac{dx \cos x}{\sin x} = dx \cot x.$$

2. Differentiate $u = \log \dfrac{a + x}{b + x}.$

By (21), $d\left(\dfrac{a + x}{b + x}\right) = \dfrac{(b + x) dx - (a + x) dx}{(b + x)^2}$

$$= -\frac{(a - b) dx}{(b + x)^2}.$$

Therefore by (25) we have

$$du = d\left(\frac{a + x}{b + x}\right) \div \frac{a + x}{b + x}$$

$$= -\frac{(a - b) dx}{(b + x)^2} \times \frac{b + x}{a + x} = -\frac{(a - b) dx}{(a + x)(b + x)}.$$

Otherwise, since $u = \log(a + x) - \log(b + x)$, we have by (25),

$$du = \frac{dx}{a + x} - \frac{dx}{b + x} = -\frac{(a - b) dx}{(a + x)(b + x)}.$$

3. Differentiate $u = e^{\sin x} \sec x.$

Here $du = \sec x \, d \left(e^{\sin x}\right) + e^{\sin x} d \sec x$, by (19),

$\qquad = \sec x \, e^{\sin x} d \sin x + e^{\sin x} d \sec x$, by (26),

$\qquad = \sec x \, e^{\sin x} . \, dx \cos x + e^{\sin x} dx \tan x \sec x$, by (29),

$\qquad = e^{\sin x} (1 + \tan x \sec x) \, dx$.

4. Differentiate $u = \log \left(\sqrt{a^2 + x^2} + x \right)$.

By (22), $d \left(\sqrt{a^2 + x^2} + x \right) = \dfrac{x \, dx}{\sqrt{a^2 + x^2}} + dx$

$$= \frac{\left(\sqrt{a^2 + x^2} + x \right) dx}{\sqrt{a^2 + x^2}}.$$

Therefore by (25) we have

$$du = \frac{d \left(\sqrt{a^2 + x^2} + x \right)}{\sqrt{a^2 + x^2} + x} = \frac{dx}{\sqrt{a^2 + x^2}}.$$

5. Differentiate $u = \log \tan e^{-x}$.

Here $du = d \left(\log \tan e^{-x} \right)$

$$= \frac{d \left(\tan e^{-x} \right)}{\tan e^{-x}}, \text{ by (25),}$$

$$= \frac{d \left(e^{-x} \right) \sec^2 e^{-x}}{\tan e^{-x}}, \text{ by (29),}$$

$$= \frac{- \, dx \, e^{-x} \left(1 + \tan^2 e^{-x} \right)}{\tan e^{-x}}, \text{ by (26),}$$

$$= - \, dx \, e^{-x} \left(\tan e^{-x} + \cot e^{-x} \right).$$

6. If $u = x^m e^{\sin x}$; then $\dfrac{du}{dx} = x^{m-1} (m + x \cos x \; e^{\sin x}$.

7. If $u = 2 \log \sin x + \operatorname{cosec}^2 x$; then $\dfrac{du}{dx} = - \, 2 \cot x$.

8. If $u = e^{\sin^{-1} x}$; then $\dfrac{du}{dx} = \dfrac{e^{\sin^{-1} x}}{\sqrt{1 - x^2}}$.

9. If $u = \log \left(\dfrac{1}{x} + \dfrac{1}{x^2} \right)$; then $\dfrac{du}{dx} = - \dfrac{x + 2}{x \, (x + 1)}$.

10. If $u = (x^2 - a^2) \log \dfrac{a + x}{a - x} + 2 a x$;

$$\text{then } \frac{du}{dx} = 2 x \log \frac{a + x}{a - x}.$$

11. If $u = \log (1 + 2 x + 2 \sqrt{1 + x + x^2})$;

$$\text{then } \frac{du}{dx} = \frac{1}{\sqrt{1 + x + x^2}}.$$

12. If $u = \tan^{-1} \dfrac{2 x}{1 - x^2}$; then $\dfrac{du}{dx} = \dfrac{2}{1 + x^2}$.

13. If $u = \sin^{-1} \dfrac{x}{\sqrt{1 + x^2}}$; then $\dfrac{du}{dx} = \dfrac{1}{1 + x^2}$.

14. If $u = \cos^{-1} \dfrac{b + a \cos x}{a + b \cos x}$; then $\dfrac{du}{dx} = \dfrac{\sqrt{a^2 - b^2}}{a + b \cos x}$.

15. If $u = \sin^{-1} (3 x - 4 x^3)$; then $\dfrac{du}{dx} = \dfrac{3}{\sqrt{1 - x^2}}$.

16. If $u = x \, e^{\tan^{-1} x}$; then $\dfrac{du}{dx} = \dfrac{1 + x + x^2}{1 + x^2} e^{\tan^{-1} x}$.

17. If $u = \tan^{-1} \sin^{-1} x$;

$$\text{then } \frac{du}{dx} = \frac{1}{\{1 + (\sin^{-1} x)^2\} \sqrt{1 - x^2}}.$$

vi. *Implicit Functions.*

(32.) The functions hitherto considered are supposed to be explicitly expressed in terms of the variable quantity involved, and upon which its value is made to depend. But a function u may have its value depending upon that of the variable x, though not expressed in any definite form, algebraical or otherwise, and perhaps not capable of being so expressed in finite terms. In fact, the relation which connects together the corresponding values of u and x may be presented in the form of an equation, $f(u, x) = 0$, f characterizing any function whatever of u and x. The function u is in such cases called an *implicit function* of the variable quantity x. If the equation $f(u, x) = 0$ could be solved for u in finite terms involving x, the function u would then be exhibited as an explicit function of x; but, as before observed, this may or may not be possible. A little consideration, however, will show that the differential of u with respect to x may be more directly obtained by taking the differential of the proposed equation in its original form.

When x becomes $x + \Delta x$, u becomes $u + \Delta u$, and as the equation $f(u, x) = 0$ must be true for all coexistent values of u and x, we have $f(u + \Delta u, x + \Delta x) = 0$, and

$$f(u + \Delta u, x + \Delta x) - f(u, x) = 0, \text{ or } \Delta f(u, x) = 0;$$

$$\therefore \frac{\Delta f(u, x)}{\Delta x} = 0.$$

This relation will be accurately true for all values of Δx, and at the limit $\Delta x = 0$ it gives

$$\frac{df(u, x)}{dx} = 0, \text{ or } df(u, x) = 0,$$

which is the differential of the proposed functional equation, observing that u and x vary simultaneously, u being a function of x. This differential equation will be of the form $P\,du + Q\,dx = 0$, and it will therefore give the value of the limiting ratio $\dfrac{du}{dx}$, or of the differential coefficient of u with respect to x, the same being expressed in terms of u and x.

Example 1.—Differentiate the function u when

$$u^2 - 2u\sqrt{a^2 + x^2} + x^2 = 0.$$

By differentiating the equation, we have

$$2u\,du - 2\sqrt{a^2 + x^2}\,du - \frac{2ux\,dx}{\sqrt{a^2 + x^2}} + 2x\,dx = 0,$$

or $2(u - \sqrt{a^2 + x^2})\,du - \dfrac{2(u - \sqrt{a^2 + x^2})}{\sqrt{a^2 + x^2}}x\,dx = 0;$

$$\therefore \frac{du}{dx} = \frac{x}{\sqrt{a^2 + x^2}}.$$

In this example the equation $u^2 - 2u \sqrt{a^2 + x^2} + x^2 = 0$ involves u in a quadratic, and may therefore be algebraically solved for u, giving $u = \sqrt{a^2 + x^2} \pm a$, which is the explicit form of the function u, and its differentiation will also lead to the result we have just obtained.

Example 2.—Differentiate u when $u^3 - 3ux^2 + 2x^3 = 0$. The differential of the equation gives

$$3u^2\,du - 3x^2\,du - 6ux\,dx + 6x^2\,dx = 0,$$

$$\text{or } 3(u^2 - x^2)\,du - 6(ux - x^2)\,dx = 0;$$

$$\therefore \frac{du}{dx} = \frac{6(ux - x^2)}{3(u^2 - x^2)} = \frac{2x}{u + x}.$$

Example 3.—Differentiate u when $x \sin u - u \sin x = 1$. By differentiating the equation, we have

$$dx \sin u + x\,du \cos u - du \sin x - u\,dx \cos x = 0,$$

$$\text{or } (x \cos u - \sin x)\,du - (u \cos x - \sin u)\,dx = 0;$$

$$\therefore \frac{du}{dx} = \frac{u \cos x - \sin u}{x \cos u - \sin x}.$$

4. If $u^3 - 3aux + x^3 = 0$; then $\dfrac{du}{dx} = -\dfrac{x^2 - au}{u^2 - ax}$.

5. If $u \sin x - x \sin u + 1 = 0$; then $\dfrac{du}{dx} = \dfrac{\sin u - u \cos x}{\sin x - x \cos u}$.

6. If $x^2 + u^2 - 2a \sqrt{x^2 - u^2} = 0$;

$$\text{then } \frac{du}{dx} = \frac{x}{u} \cdot \frac{a - \sqrt{x^2 - u^2}}{a + \sqrt{x^2 - u^2}}.$$

7. If $u^n \log u - ax = 0$; then $\dfrac{du}{dx} = \dfrac{a}{u^{n-1}(1 + n \log u)}$.

8. If $x e^u - u + 1 = 0$; then $\dfrac{du}{dx} = \dfrac{e^u}{2 - u}$.

9. If $ux - (a + u) \sqrt{b^2 - u^2} = 0$;

$$\text{then } \frac{du}{dx} = -\frac{u}{x} \cdot \frac{(a + u)(b^2 - u^2)}{ab^2 + u^3}.$$

10. If $\log \dfrac{a + \sqrt{a^2 - u^2}}{u} - \dfrac{x + \sqrt{a^2 - u^2}}{a} = 0$;

then $\dfrac{du}{dx} = - \dfrac{u}{\sqrt{a^2 - u^2}}.$

VII. *Functions of Two or more Variables.*

(33.) Let $u = f(x, y)$ denote a function of two variables x and y.

If instead of x and y varying simultaneously, x be supposed to vary alone without any change in the value of y, then y will be treated as the symbol of a constant quantity, and u being then considered as a function of x only, its differential or differential coefficient will be determined by the foregoing methods for functions of one variable. The value so determined, however, as it is made to depend upon a change in the value of x without any supposed change in the value of y, will be only partial, and will not refer to a consideration of the total change of u. In order to distinguish this, the differential coefficient is usually placed within a parenthesis; thus $\left(\dfrac{du}{dx}\right)$ denotes the *partial* differential coefficient, and $\left(\dfrac{du}{dx}\right) dx$ the *partial* differential of u *with respect to x*, that is, supposing x alone to change. Similarly, if y alone be supposed to vary and x to be invariable, $\left(\dfrac{du}{dy}\right)$ will denote the *partial* differential coefficient, and $\left(\dfrac{du}{dy}\right) dy$ the *partial* differential of u *with respect to y*. These partial differentiations, as before observed, may be effected by the preceding methods for functions of a single variable; first regarding u as a function of only one variable x, and again as a function of only one variable y.

The supposition of x or y varying separately, so as to partially differentiate the function u, is here to be received as a mere conventional hypothesis assumed for the purpose of more distinctly defining certain abstract analytical operations, to be applied hereafter.

Returning now to the proposed function $u = f(x, y)$, when x and y respectively become $x + \Delta x$, $y + \Delta y$, it becomes

$$u + \Delta u = f(x + \Delta x, y + \Delta y) ;$$

that $\Delta u = f(x + \Delta x, y + \Delta y) - f(x, y)$, which denotes the otal increment of u, or the combined effect produced on the alue of the function by the two increments Δx, Δy. Instead of conceiving the values of x and y to change simultaneously, we may suppose them to change successively, as the result will be the same in both cases.

Thus, supposing x to become $x + \Delta x$ and the value of y to remain unchanged, the function $f(x, y)$ will become

$$f(x + \Delta x, y) ;$$

and again, supposing, in this last function, y to become $y + \Delta y$ and x to remain unchanged, it will become $f(x + \Delta x, y + \Delta y)$, which is the complete value of u consequent on the changes in the values of x and y. The function u instead of passing at once to this last value is made to assume the three values $f(x, y)$, $f(x + \Delta x, y)$, $f(x + \Delta x, y + \Delta y)$, and the partial increments of u in successively passing to these values are,

$$f(x + \Delta x, y) - f(x, y)$$
$$= \Delta f(x, y) \text{ with respect to } x ;$$
$$f(x + \Delta x, y + \Delta y) - f(x + \Delta x, y)$$
$$= \Delta f(x + \Delta x, y) \text{ with respect to } y :$$

the sum of which gives $f(x + \Delta x, y + \Delta y) - f(x, y) = \Delta u$, the total increment of u.

$$\therefore \frac{\Delta u}{\Delta x} = \frac{\Delta f(x, y) \text{ with respect to } x}{\Delta x}$$
$$+ \frac{\Delta f(x + \Delta x, y) \text{ with respect to } y}{\Delta y} \cdot \frac{\Delta y}{\Delta x}.$$

Hence, taking the limiting values when $\Delta x = 0$, $\Delta y = 0$, we obtain

$$\frac{du}{dx} = \left(\frac{du}{dx}\right) + \left(\frac{du}{dy}\right)\frac{dy}{dx}.$$

$$\therefore \ du = \left(\frac{du}{dx}\right)dx + \left(\frac{du}{dy}\right)dy.$$

The differential of a function of two variables is therefore found by taking the sum of the partial differentials.

(34.) Again, let $u = f(x, y, z)$ be a function involving three variables x, y, and z; then

$$\Delta u = f(x + \Delta x, y + \Delta y, z + \Delta z) - f(x, y, z).$$

But, instead of considering the values of x, y, z to change simultaneously, we may, as before, suppose them to change successively. In this way the function u, instead of passing at once to the new value $f(x + \Delta x, y + \Delta y, z + \Delta z)$, will be made to assume the four values $f(x, y, z)$, $f(x + \Delta x, y, z)$,

$$f(x + \Delta x, y + \Delta y, z), f(x + \Delta x, y + \Delta y, z + \Delta z),$$

and the partial increments of u in successively passing to these values will be

$$f(x + \Delta x, y, z) - f(x, y, z)$$
$$= \Delta f(x, y, z) \text{ with respect to } x;$$

$$f(x + \Delta x, y + \Delta y, z) - f(x + \Delta x, y, z)$$
$$= \Delta f(x + \Delta x, y, z) \text{ with respect to } y;$$

$$f(x + \Delta x, y + \Delta y, z + \Delta z) - f(x + \Delta x, y + \Delta y, z)$$
$$= \Delta f(x + \Delta x, y + \Delta y, z) \text{ with respect to } z:$$

the sum of which gives

$$f(x + \Delta x, y + \Delta y, z + \Delta z) - f(x, y, z) = \Delta u,$$

the total increment of u.

$$\therefore \ \frac{\Delta u}{\Delta x} = \frac{\Delta f(x, y, z) \text{ with respect to } x}{\Delta x}$$

$$+ \frac{\Delta f(x + \Delta x, y, z) \text{ with respect to } y}{\Delta y} \cdot \frac{\Delta y}{\Delta x}$$

$$+ \frac{\Delta f(x + \Delta x, y + \Delta y, z) \text{ with respect to } z}{\Delta z} \cdot \frac{\Delta z}{\Delta x}.$$

Hence, proceeding to the limiting values when $\Delta x = 0$, $\Delta y = 0$, $\Delta z = 0$, we have

$$\frac{du}{dx} = \left(\frac{du}{dx}\right) + \left(\frac{du}{dy}\right)\frac{dy}{dx} + \left(\frac{du}{dz}\right)\frac{dz}{dx};$$

$$\therefore \ du = \left(\frac{du}{dx}\right) dx + \left(\frac{du}{dy}\right) dy + \left(\frac{du}{dz}\right) dz.$$

The differential of a function of three variables is therefore obtained by taking the sum of the partial differentials; and this principle evidently extends to functions of any number of variables.

Example 1.—If $u = x \log y$; then supposing x only to vary we have

$$\left(\frac{du}{dx}\right) = \log y \ ; \ \text{and supposing } y \text{ only to vary, } \left(\frac{du}{dy}\right) = \frac{x}{y};$$

$$\therefore \ du = (\log y) \, dx + \left(\frac{x}{y}\right) dy.$$

Example 2.—If $u = x^3 + 3\,a\,x\,y + y^3$;

$$\text{then } \left(\frac{du}{dx}\right) = 3 (x^2 + a\,y), \ \left(\frac{du}{dy}\right) = 3 (y^2 + a\,x);$$

$$\therefore \ du = 3 (x^2 + a\,y) \, dx + 3 (y^2 + a\,x) \, dy.$$

Example 3.—If $u = \dfrac{x - y}{x + y}$:

$$\text{then } \left(\frac{du}{dx}\right) = \frac{2\,y}{(x + y)^2}, \ \left(\frac{du}{dy}\right) = -\frac{2\,x}{(x + y)^2};$$

$$\therefore \ du = \frac{2 (y \, dx - x \, dy)}{(x + y)^2}.$$

c

4. If $u = x + y + \sqrt{x^2 + y^2}$;

then $du = \left(1 + \dfrac{x}{\sqrt{x^2 + y^2}}\right) dx + \left(1 + \dfrac{y}{\sqrt{x^2 + y^2}}\right) dy.$

5. If $u = x^y$; then $du = (y\, x^{y-1})\, dx + (x^y \log x)\, dy.$

6. If $u = x y \sqrt{x^2 + y^2}$;

then $du = \dfrac{(2\,x^2 + y^2)\, y\, dx + (x^2 + 2\,y^2)\, x\, dy}{\sqrt{x^2 + y^2}}.$

7. If $u = \dfrac{x^m}{y^n}$; then $du = \dfrac{x^{m-1}}{y^{n+1}} (m\, y\, dx - n\, x\, dy).$

8. If $u = \cos x \sin y + \sin x \cos y$;

then $du = (dx + dy)(\cos x \cos y - \sin x \sin y).$

9. If $u = x \sqrt{a^2 + y^2} + y \sqrt{b^2 - x^2}$; then

$$du = \left(\sqrt{a^2 + y^2} - \dfrac{x y}{\sqrt{b^2 - x^2}}\right) dx$$

$$+ \left(\dfrac{x y}{\sqrt{a^2 + y^2}} + \sqrt{b^2 - x^2}\right) dy.$$

10. If $u = x y z$; then $du = y z\, dx + z x\, dy + x y\, dz.$

11. If $u = x y + y z + z x$;

then $du = (y + z)\, dx + (z + x)\, dy + (x + y)\, dz.$

12. If $u = \dfrac{\sqrt{x^2 + y^2 + z^2}}{x y z}$;

then $du = - \dfrac{(y^2 + z^2)\dfrac{dx}{x} + (z^2 + x^2)\dfrac{dy}{y} + (x^2 + y^2)\dfrac{dz}{z}}{x y z \sqrt{x^2 + y^2 + z^2}}$

13. If $u = \dfrac{y - z}{z - x}$;

then $du = \dfrac{(y - z)\, dx + (z - x)\, dy + (x - y)\, dz}{(z - x)^2}.$

CHAPTER III.

SUCCESSIVE DIFFERENTIATION.

1. *Functions of One Variable.*

(35.) By differentiating a function $u = fx$, of a variable quantity x, it has been shown that the differential coefficient $\dfrac{du}{dx}$ will be another function $f'x$, and the methods of determining it have been established in the last Chapter. By similarly differentiating this new function $f'x$ so as to obtain its differential coefficient denoted by $f''x$, this is called the second differential coefficient of the original function fx. In like manner if we differentiate $f''x$, its differential coefficient $f'''x$ is called the third differential coefficient of the function fx; and, provided the variable quantity x does not disappear from these functions, this operation may evidently be repeated to any order of differentiation. This continued process is called *successive differentiation*, and it is indicated by the following relations :

$$f'x = \frac{du}{dx}, \quad f''x = \frac{d\left(\dfrac{du}{dx}\right)}{dx}, \quad f'''x = \frac{d\left\{\dfrac{d\left(\dfrac{du}{dx}\right)}{dx}\right\}}{dx}, \ \&c.$$

which may also be thus expressed,

$$f'x = \frac{du}{dx}, \quad f''x = \frac{d}{dx}\frac{du}{dx}, \quad f'''x = \frac{d}{dx}\frac{d}{dx}\frac{du}{dx}, \ \&c.$$

According to Lagrange, fx is the primitive function, and $f'x$, $f''x$, $f'''x$, &c., thus determined, are respectively called the first, second, third, &c. derived functions.—See art. (11).

Although in the original idea of differentiation as founded on the theory of limits, a differential can have only a relative signification, yet, when separately considered as an infinitesimal change of the variable, it may in analytical calculations be regarded and operated upon as an indeterminate quantity, the value of which is only appreciable when it is compared with other quantities of the same order or kind.

Thus the differentiation $f'x = \dfrac{du}{dx}$ merely defines the value of the ultimate ratio of two infinitesimal elements du and dx, and, in other respects, we are at liberty to assign any law whatever to the separate values of these elements as depending upon x. We might suppose the values of du and dx to be each of them different for different values of x, so as to change when x changes. It will, however, conveniently simplify our notation if x be taken as an independent variable; that is, if we suppose the infinitesimal increment dx to have the same fixed value for all values of x, so as to admit of being treated as a constant. In this case x is tacitly supposed to increase by equal infinitesimal increments dx, and dx is thus independent of the value of x; but the value of $du = dx f'x$ will evidently depend upon that of x and be different for different values of x. Hence the reason why x is in such case specially called the *independent variable*; also as the invariable element dx is to be regarded as a constant in each differentiation, the foregoing relations obviously become

$$f'x = \frac{du}{dx}, \quad f''x = \frac{d\,(du)}{dx^2}, \quad f'''x = \frac{d\,\{d\,(du)\}}{dx^3}, \ \&c.$$

Or, in accordance with the general index law, these are more conveniently written

$$f'x = \frac{du}{dx}, \quad f''x = \frac{d^2u}{dx^2}, \quad f'''x = \frac{d^3u}{dx^3}, \ \&c.$$

And thus the symbols $\dfrac{du}{dx}, \ \dfrac{d^2u}{dx^2}, \ \dfrac{d^3u}{dx^3}$, &c. represent the first, second, third, &c. differential coefficients of u with respect to x; or separately considering the numerators and denominators, du, d^2u, d^3u, &c. denote the first, second, third, &c. differentials of u supposing dx to be constant, and dx, dx^2, dx^3, &c. as before, indicate dx, $(dx)^2$, $(dx)^3$, &c. or powers of dx.

Example 1. Let $u = x^n$; then $\dfrac{du}{dx} = n\,x^{n-1}$,

$$\frac{d^2u}{dx^2} = n\,(n-1)\,x^{n-2}, \quad \frac{d^3u}{dx^3} = n\,(n-1)\,(n-2)\,x^{n-3}, \text{ \&c.,}$$

$$\frac{d^n u}{dx^n} = n\,(n-1)\,(n-2)\,(n-3)\ldots\ldots 1 = 1.2.3\ldots.n.$$

Ex. 2. Let $u = e^x$; then by (26),

$$\frac{du}{dx} = e^x, \quad \frac{d^2u}{dx^2} = e^x, \ldots\ldots \frac{d^n u}{dx^n} = e^x.$$

Ex. 3. Let $u = \cos x$; then $\dfrac{du}{dx} = -\sin x = \cos\left(x + \dfrac{\pi}{2}\right)$,

$$\frac{d^2u}{dx^2} = -\cos x = \cos\left(x + \frac{2\pi}{2}\right),$$

$$\frac{d^3u}{dx^3} = \sin x = \cos\left(x + \frac{3\pi}{2}\right), \text{ \&c.} \ldots\ldots,$$

$$\frac{d^n u}{dx^n} = \cos\left(x + \frac{n\pi}{2}\right).$$

Ex. 4. Let $u = e^x \cos x$; then

$$\frac{du}{dx} = e^x \cos x - e^x \sin x = e^x(\cos x - \sin x)$$

$$= \sqrt{2}\,e^x \cos\left(x + \frac{\pi}{4}\right)$$

$$\frac{d^2u}{dx^2} = \sqrt{2}\,e^x \left\{ \cos\left(x + \frac{\pi}{4}\right) - \sin\left(x + \frac{\pi}{4}\right) \right\}$$

$$= (\sqrt{2})^2\,e^x \cos\left(x + \frac{2\pi}{4}\right)$$

$$\text{\&c. \quad \&c.}$$

$$\frac{d^n u}{dx^n} = (\sqrt{2})^n\,e^x \cos\left(x + \frac{n\pi}{4}\right).$$

Ex. 5. If $u = x^3 + a x^2 + b x + c$; then $\dfrac{d^3 u}{dx^3} = 1.2.3$.

Ex. 6. If $u = \sin x$; then $\dfrac{d^n u}{dx^n} = \sin\left(x + \dfrac{n\pi}{2}\right)$.

Ex. 7. If $u = e^{mx}$; then $\dfrac{d^n u}{dx^n} = m^n e^{mx}$.

Ex. 8. If $u = x e^x$; then $\dfrac{d^n u}{dx^n} = (x + n) e^x$.

Ex. 9. If $u = e^x \sin x$;

then $\dfrac{d^n u}{dx^n} = (\sqrt{2})^n e^x \sin\left(x + \dfrac{n\pi}{4}\right)$.

Ex. 10. If $u = \dfrac{1 + x}{1 - x}$; then $\dfrac{d^n u}{dx^n} = \dfrac{1.2.3 \ldots . n}{(1 - x)^{n+1}}$.

II. *Changing of the Independent Variable.*

(36.) When an expression involving two variables x, y and the successive differential coefficients has been arrived at on the supposition that one of the variables is independent, it is sometimes required to transform it into its equivalent when the other variable is independent. This process is called changing the independent variable, and it is accomplished by replacing the second and higher differential coefficients by their complete values supposing no independent variable to be assumed, and afterwards introducing whatever new condition may be necessary.

Thus if $\dfrac{d^2 y}{dx^2}$, $\dfrac{d^3 y}{dx^3}$, &c. have been calculated with respect to x as an independent variable, to replace these coefficients by the general values when x is not independent, and therefore dx not constant, we shall have, art. (21),

$$\frac{d^2 y}{dx^2} = \frac{d\left(\dfrac{dy}{dx}\right)}{dx} = \frac{d^2 y \, dx - d^2 x \, dy}{dx^3},$$

$$\frac{d^3y}{dx^3} = \frac{d\left(\frac{d^2y}{dx^2}\right)}{dx}$$

$$= \frac{(d^3y\,dx - d^3x\,dy)\,dx - 3\,(d^2y\,dx - d^2x\,dy)\,d^2x}{dx^5},$$

&c. &c. &c.

By substituting these values in place of $\dfrac{d^2y}{dx^2}$, $\dfrac{d^3y}{dx^3}$, &c. we shall obtain the corresponding expression when neither x nor y is supposed to be an independent variable. If y is required to be an independent variable in the new expression, we must make $d^2y = 0$, $d^3y = 0$, &c., in which case the equivalents will be

$$\frac{d^2y}{dx^2} = -\frac{d^2x\,dy}{dx^3},$$

$$\frac{d^3y}{dx^3} = \frac{3\,(d^2x)^2\,dy - d^3x\,dy\,dx}{dx^5},$$

&c. &c.

by the substitution of which the independent variable will be at once changed from x to y.

III. *Functions of Two or more Variables.*

(37.) In art. (33) it has been shown that the total differential of a function of two variables is obtained by taking the sum of the partial differentials, supposing each of them to vary alone. That is, if $u = f(x, y)$, we have

$$du = \left(\frac{du}{dx}\right) dx + \left(\frac{du}{dy}\right) dy.$$

As the partial differential coefficients $\left(\dfrac{du}{dx}\right)$, $\left(\dfrac{du}{dy}\right)$ are also functions of the two variables x, y, it is evident that the value of du will admit of being differentiated again in a similar manner so as to obtain d^2u, and that this operation may be repeated up to any required order of differentiation. To exhibit the results of these processes it will be requisite to extend our notation.

When a function u is successively differentiated with respect to x, considered as an independent variable, the results, according to the notation of art. (35), are thus indicated,

$$\left(\frac{du}{dx}\right), \quad \left(\frac{d^2u}{dx^2}\right), \quad \left(\frac{d^3u}{dx^3}\right), \quad \&c. \ \&c.$$

The same with respect to y are

$$\left(\frac{du}{dy}\right), \quad \left(\frac{d^2u}{dy^2}\right), \quad \left(\frac{d^3u}{dy^3}\right), \quad \&c. \ \&c.,$$

the brackets indicating, as in art. (33), that the derived functions are only partial.

But we may differentiate, in succession, sometimes with respect to one variable and sometimes another, in which cases the notation usually adopted is as follows :

$$\frac{d}{dx}\left(\frac{du}{dy}\right) \text{ is indicated by } \left(\frac{d^2u}{dx\,dy}\right)$$

$$\frac{d}{dx}\frac{d}{dx}\left(\frac{du}{dy}\right) \text{ is indicated by } \left(\frac{d^3u}{dx^2\,dy}\right),$$

$$\&c. \qquad\qquad\qquad \&c.$$

where the numerator shows how many differentiations have been taken, and the denominator shows the variables employed in the reverse order of the operations. We proceed to show that the resulting values of these successive partial derived functions are independent of the order in which the variables are supposed to change.

The operation of differentiating a function $\phi(x)$ is defined by the relation

$$\frac{d\phi(x)}{dx} \text{ or } \frac{d}{dx}\phi(x) = \frac{\phi(x + dx) - \phi(x)}{dx}.$$

By applying this to the function $u = f(x, y)$, first with respect to x and then with respect to y, we have

$$\left(\frac{du}{dx}\right) = \frac{f(x + dx, y) - f(x, y)}{dx},$$

$$\left(\frac{du}{dy}\right) = \frac{f(x, y + dy) - f(x, y)}{dy};$$

and by again applying the same principle to these functions, we get

$$\frac{d}{dy}\left(\frac{du}{dx}\right) =$$

$$\frac{f(x + dx, y + dy) - f(x, y + dy) - f(x + dx, y) + f(x, y)}{dx\,dy}$$

$$\frac{d}{dx}\left(\frac{du}{dy}\right) =$$

$$\frac{f(x + dx, y + dy) - f(x + dx, y) - f(x, y + dy) + f(x, y)}{dx\,dy}.$$

Hence, as these expressions are alike, we have

$$\frac{d}{dy}\left(\frac{du}{dx}\right) = \frac{d}{dx}\left(\frac{du}{dy}\right);$$

that is,

$$\left(\frac{d^2 u}{dy\,dx}\right) = \left(\frac{d^2 u}{dx\,dy}\right).$$

This property is true when u is a function of any number of variables, because when x and y alone vary, the other variables only enter in the same manner as constants, and as regards the operations performed, u may therefore be considered as a function of only two variables. Hence it follows that in calculating partial differential coefficients we may always interchange at pleasure the order in which the several differentiations are performed, without altering the results. Thus when $u = f(x, y)$, we have also

$$\left(\frac{d^3 u}{dy\,dx^2}\right) = \left(\frac{d^3 u}{dx^2\,dy}\right), \qquad \left(\frac{d^3 u}{dy^2\,dx}\right) = \left(\frac{d^3 u}{dx\,dy^2}\right);$$

and generally, when u is a function of two variables,

$$\left(\frac{d^{r+s}u}{dy^s\,dx^r}\right) = \left(\frac{d^{r+s}u}{dx^r\,dy^s}\right),$$

$$\frac{d}{dx}\left(\frac{d^{r+s}u}{dx^r\,dy^s}\right) = \left(\frac{d^{r+s+1}u}{dx^{r+1}\,dy^s}\right), \quad \frac{d}{dy}\left(\frac{d^{r+s}u}{dx^r\,dy^s}\right) = \left(\frac{d^{r+s+1}u}{dx^r\,dy^{s+1}}\right).$$

Example 1. Let $u = x \sin y + y \sin x$; then

$$\left(\frac{du}{dx}\right) = \sin y + y \cos x, \quad \left(\frac{du}{dy}\right) = x \cos y + \sin x;$$

$$\therefore \; \left(\frac{d^2u}{dy\,dx}\right) = \cos y + \cos x, \quad \left(\frac{d^2u}{dx\,dy}\right) = \cos y + \cos x,$$

which two results are identical.

Ex. 2. Let $u = 2x^2y^3 + x^4y$; then

$$\left(\frac{du}{dy\,dx\,dx}\right) = \left(\frac{du}{dx\,dy\,dx}\right) = \left(\frac{du}{dx\,dx\,dy}\right) = 12(x^2 + y^2).$$

(38.) The general property established in the last article will assist us in the successive differentiation of a function of two or more variables. Let $u = f(x, y)$, a function of two variables; then, art. (33), its first complete differential is

$$du = \left(\frac{du}{dx}\right) dx + \left(\frac{du}{dy}\right) dy.$$

In proceeding to the next differentiation it must be observed that the coefficients $\left(\dfrac{du}{dx}\right)$, $\left(\dfrac{du}{dy}\right)$ are generally to be considered as functions of both variables, and to separately admit of being differentiated in the same manner as the original function u, by adding together the partial differentials. Thus we have

$$d\left(\frac{du}{dx}\right) = \frac{d}{dx}\left(\frac{du}{dx}\right) dx + \frac{d}{dy}\left(\frac{du}{dx}\right) dy$$

$$= \left(\frac{d^2u}{dx\,dy}\right) dx + \left(\frac{d^2u}{dy^2}\right) dy.$$

Again, if we adopt the principle of general differentiation, and suppose dx and dy to be variable, we shall have, art. (19),

$$d\left\{\left(\frac{du}{dx}\right) dx \right\} = dx \cdot d\left(\frac{du}{dx}\right) + \left(\frac{du}{dx}\right) d^2x$$

$$d\left\{\left(\frac{du}{dy}\right) dy \right\} = dy \cdot d\left(\frac{du}{dy}\right) + \left(\frac{du}{dy}\right) d^2y.$$

The sum of the left-hand members of these is the differential of the value of du, and is therefore equal to d^2u. Hence, adding together these two equations and substituting the preceding values of $d\left(\frac{du}{dx}\right)$, $d\left(\frac{du}{dy}\right)$, we obtain

$$d^2u = \left(\frac{d^2u}{dx^2}\right) dx^2 + 2\left(\frac{d^2u}{dx\,dy}\right) dx\,dy + \left(\frac{d^2u}{dy^2}\right) dy^2$$

$$+ \left(\frac{du}{dx}\right) d^2x + \left(\frac{du}{dy}\right) d^2y.$$

The process of differentiation may be successively carried on to higher orders in precisely the same manner, so as to determine general expressions for d^3u, d^4u, &c.; but as the formulæ for the higher orders become rather cumbrous and are seldom required, it will not be necessary to give any of them here.

If the variables x and y are independent of each other, and their values admit of being connected by a relation of the form $y = ax + \beta$, so that we may consider both of them to increase by constant increments; then dx and $dy = a\,dx$ may be both supposed to be invariable. On this hypothesis, $d^2x = 0$, &c. and $d^2y = 0$, &c. and the expressions become

$$u = f(x, y),$$

$$du = \left(\frac{du}{dx}\right) dx + \left(\frac{du}{dy}\right) dy,$$

$$d^2u = \left(\frac{d^2u}{dx^2}\right) dx^2 + 2\left(\frac{d^2u}{dx\,dy}\right) dx\,dy + \left(\frac{d^2u}{dy^2}\right) dy^2,$$

&c.　　　　&c.　　　　&c.

Here the numerical coefficients will be found to observe the same law as those of the binomial theorem; and the nth differential may be put down as follows :

$$d^n u = \left(\frac{d^n u}{dx^n}\right) dx^n + n \left(\frac{d^n u}{dx^{n-1}dy}\right) dx^{n-1}dy$$

$$+ \frac{n(n-1)}{1.2} \left(\frac{d^n u}{dx^{n-2}dy^2}\right) dx^{n-2}dy^2 \ldots + \left(\frac{d^n u}{dy^n}\right) dy^n.$$

The successive differentiations of a function of any number of variables may be determined in the same way as the preceding. Let $u = f(x, y, z)$ be a function of three independent variables, and suppose $y = ax + \beta$, $z = a'x + \beta'$, so that x, y and z may severally increase by constant increments; then we find

$$u = f(x, y, z),$$

$$du = \left(\frac{du}{dx}\right) dx + \left(\frac{du}{dy}\right) dy + \left(\frac{du}{dz}\right) dz,$$

$$d^2 u = \left(\frac{d^2 u}{dx^2}\right) dx^2 + \left(\frac{d^2 u}{dy^2}\right) dy^2 + \left(\frac{d^2 u}{dz^2}\right) dz^2$$

$$+ 2 \left(\frac{d^2 u}{dy\, dz}\right) dy\, dz + 2 \left(\frac{d^2 u}{dz\, dx}\right) dz\, dx + 2 \left(\frac{d^2 u}{dx\, dy}\right) dx\, dy,$$

&c.   &c.   &c.

CHAPTER IV.

EXPANSION OF FUNCTIONS.

1. *Functions of One Variable.*

(39.) Let $u = f(x)$ denote a function of x, and, h denoting a finite quantity, let the binomial function $f(x + h)$ when expanded in terms involving the integral powers of h be supposed to be

$$f(x + h) = f(x) + Ph + Qh^2 + Rh^3 + \&c.,$$

in which P, Q, R,, &c. are new functions of x to be determined from $f(x)$. It has been shown, art. (6), that the coefficient P of the second term of this development is the differential coefficient of the function $f(x)$, and is therefore to be obtained at once by differentiation. The other coefficients Q, R, &c. may be similarly determined by means of successive differentiation. Thus, by differentiating successively the above form of expansion, we get the following equations :

$$f'(x + h) = P + 2\,Q\,h + \quad 3\,R\,h^2 + \&c.$$

$$f''(x + h) = \quad 1.2\,Q \quad + 2.3\,R\,h \; + \&c.$$

$$f'''(x + h) = \quad\quad\quad 1.2.3\,R \quad + \&c.$$

$$\&c. \quad\quad\quad\quad\quad \&c.$$

As these must be true for all values of h, by supposing the coefficients P, Q, R, &c. to be finite in value, and making $h = 0$, we obtain,

$$f'(x) = P, \quad f''(x) = 1.2\,Q, \quad f'''(x) = 1.2.3\,R, \&c. \&c. ;$$

$$\therefore P = \frac{f'(x)}{1}, \quad Q = \frac{f''(x)}{1.2}, \quad R = \frac{f'''(x)}{1.2.3}, \&c. \quad \&c.$$

Hence the expansion of $f(x + h)$ is,

$$f(x + h) = f(x) + f'(x)\,\frac{h}{1} + f''(x)\,\frac{h^2}{1.2} + f'''(x)\,\frac{h^3}{1.2.3} + \&c.$$

$$= u + \frac{du}{dx} \cdot \frac{h}{1} + \frac{d^2u}{dx^2} \cdot \frac{h^2}{1.2} + \frac{d^3u}{dx^3} \cdot \frac{h^3}{1.2.3} + \&c.,$$

which is Taylor's theorem, and is one of considerable importance.

In deducing it we have in the first place assumed without proof that the function is capable of being developed in the proposed form. The mere fact of obtaining an intelligible result will, however, be sufficient to establish the truth of this supposition.

We have also necessarily assumed that *all* the coefficients

P, Q, R, &c. should be finite, as the reasoning evidently ceases to be conclusive when any of these coefficients become infinite in value.   When one of these coefficients becomes infinite in value, we shall find that all the coefficients which succeed it will also be infinite in value.   Whenever this happens, which can only be in particular cases and for particular values of x, Taylor's theorem is commonly said to *fail;* but it may in such cases be more properly said to be *inapplicable,* in conse-quence of the impossibility of exhibiting the complete expansion of the given function in the required form for that particular value of x.   We shall hereafter give a more satisfactory investigation of the development in a modified form, so as to obviate any want of generality or of logical accuracy that would otherwise be experienced in the many important applications of this celebrated theorem

(40.) By making $x = 0$, Taylor's theorem becomes

$$f(h) = f(0) + f'(0) \frac{h}{1} + f''(0) \frac{h^2}{1.2} + f'''(0) \frac{h^3}{1.2.3} + \&c.$$

Or, substituting x for h,

$$f(x) = f(0) + f'(0) \frac{x}{1} + f''(0) \frac{x^2}{1.2} + f'''(0) \frac{x^3}{1.2.3} + \&c.,$$

which is generally known as "Maclaurin's theorem," and is useful for the expansion of functions in powers of the variable. Professor De Morgan has observed, that Maclaurin was anticipated in the use of this theorem, and it has in consequence been latterly called "Stirling's theorem ;" but of this it may be remarked, that it is an obvious and very easily deduced particular case of Taylor's theorem, of still earlier date ; being in fact, merely the development of $f(x)$ considered as a binomial function $f(0 + x)$.

II. *Theorems which Limit the Values of Functions.*

(41.) Let $f(x)$, $f(x + h)$ be two values of a function which varies continuously between x and $x + h$; then if any value of

x between x and $x + h$ be substituted in the proposed function, the result will be an *intermediate function.* For example, the functions $f(x + \frac{1}{2}h), f(x + \frac{1}{3}h), f\left(x + \dfrac{m}{m + n}\,h\right)$ are all intermediate functions with respect to $f(x)$ and $f(x + h)$; but it does not necessarily follow that their values are arithmetically intermediate between $f(x)$ and $f(x + h)$ unless the function between these limits either continually increases or continually decreases. If, however, x be supposed to vary continuously and to take every possible value from x to $x + h$, and V, v denote respectively the greatest and least values of the function between those limits, then the value of every intermediate function will obviously be comprised between V and v.

(42.) When a variable x takes m progressive values x_1, x_2, $x_3 \ldots\ldots x_m$, let the corresponding values of a function $u = f(x)$ be denoted by $u_1, u_2, u_3 \ldots\ldots u_m$; then if the function be continuous in value from u_1 to u_m we shall have

$$u_1 + u_2 + u_3 \ldots\ldots + u_m = m\,u_{\theta m}$$

where θ is some arithmetical value between zero and unity, so that the value of θm is between 1 and m, and $u_{\theta m}$ is a function of x intermediate with respect to u_1 and u_m.

Let V, v denote the greatest and least values of the function u when x is supposed to pass continuously through every value from x_1 to x_m, so that $u_1, u_2, u_3 \ldots\ldots u_m$ are severally comprised between them, that is, less than V and greater than v; also let the sum of these m functions be denoted by $m\,(u)$, then

$$V + V + V \ \&c.\ \text{to } m \text{ terms} = m\,V \ \ldots\ldots (1)$$
$$u_1 + u_2 + u_3 \ldots\ldots + u_m = m\,(u) \ \ldots\ldots (2)$$
$$v + v + v \ \&c.\ \text{to } m \text{ terms} = m\,v \ \ldots\ldots (3).$$

On inspecting these we observe that the terms of (2) are severally less than the corresponding terms of (1) and greater than the corresponding terms of (3), and therefore the total value of (2) is less than that of (1) and greater than that of (3). That is, the value of (u) is comprised between V and v, and is therefore a value of the function between these values. Hence, as V and v are each intermediate with respect to u_1 and u_m, (u) must necessarily be the value of an intermediate function with respect to u_1 and u_m, and may therefore be represented by $u_{\theta m}$, θ expressing a numerical value between zero and unity.

It will be observed that the basis of this proof is the evident proposition that when, with respect to certain functional limits, a value is arithmetically intermediate it must also be functionally intermediate, provided that the function is continuous between the stated limits.

(43.) Let $f(x)$ be a function of x, continuous and finite from 0 to x, and which vanishes when $x = 0$; then will

$$f(x) = x f'(\theta x),$$

where θ is some arithmetical value between zero and unity.

Suppose x to be divided into a number (m) of parts, each equal to dx, so that $m \, dx = x$, the number m being indefinitely great and dx indefinitely small. Then, according to the first principle of differentiation,

$$\frac{f(0 + dx) - f(0)}{dx} = f'(0)$$

$$\frac{f(dx + dx) - f(dx)}{dx} = f'(dx)$$

$$\frac{f(2\,dx + dx) - f(2\,dx)}{dx} = f'(2\,dx)$$

$$\frac{f(3\,dx + dx) - f(3\,dx)}{dx} = f'(3\,dx)$$

$$\&c. \qquad\qquad \&c.$$

$$\frac{f(m\,dx) - f\{(m-1)\,dx\}}{dx} = f'\{(m-1)\,dx\}.$$

Hence, observing that $m \, dx = x$, the sum of these equations, according to (42), gives

$$\frac{f(x) - f(0)}{} = m f'(\theta x),$$

or, since $f(0) = 0$,

$$f(x) = m \, dx f'(\theta x) = x f'(\theta x).$$

Cor. If a function $f(x)$ be continuous in value from 0 to x, and also vanishes at each of these limits, so that $f(0) = 0$, $f(x) = 0$; then, by the preceding theorem,

$$x f'(\theta x) = f(x) = 0;$$

$$\therefore \ f'(\theta x) = 0.$$

That is, if $f(x)$ vanishes at both of the values 0 and x, the derived function or differential coefficient $f'(x)$ will vanish at θx, some value between 0 and x.

(44.) If $f(h)$ a function of h together with its first n derived functions be finite and continuous from 0 to h; and if moreover the function and the first $n - 1$ of these derived functions severally vanish when $h = 0$; then

$$f(h) = \frac{h^n}{1.2.3 \ldots n} f^{(n)}(\theta h),$$

where θ is some positive arithmetical value less than unity.

Let h be supposed to be constant and x variable, and assume

$$F(x) = h^n f(x) - x^n f(h).$$

Then, since $F(x)$ vanishes when $x = 0$ and $x = h$, it follows from the corollary to (43), that the derived function

$$F'(x) = h^n f'(x) - n x^{n-1} f(h)$$

will vanish when $x = \theta_1 h = h_1$, where h_1 is some value between 0 and h. But since, by hypothesis, $f'(0) = 0$, this derived function $F'(x)$ also vanishes when $x = 0$. Hence again, as the function $F'(x)$ vanishes when $x = 0$ and $x = h_1$, it follows from the same corollary, that its derived function

$$F''(x) = h^n f''(x) - n(n-1) x^{n-2} f(h)$$

will vanish when $x = h_2$, some value between 0 and h_1. But since, by hypothesis, $f''(0) = 0$, this function $F''(x)$ also vanishes when $x = 0$. Hence, as before,

$$F'''(x) = h^n f'''(x) - n(n-1)(n-2) x^{n-3} f(h)$$

will vanish when $x = h_3$, some value between 0 and h_2.

By pursuing this process we shall evidently find that

$$F^{(n)}(x) = h^n f^{(n)}(x) - n(n-1)(n-2) \dots 1 f(h)$$

vanishes when $x = h_n$, some value between 0 and h_{n-1}. That is, substituting for x this last value,

$$h^n f^{(n)}(h_n) - 1.2.3 \dots n f(h) = 0;$$

$$\therefore f(h) = \frac{h^n}{1.2.3 \dots n} f^{(n)}(h_n),$$

where h_n is some value between 0 and h, which may therefore be designated by θh, θ being an arithmetical value between zero and unity. Hence we have

$$f(h) = \frac{h^n}{1.2.3 \dots n} f^{(n)}(\theta h),$$

which is a further extension of the theorem of art. (43).

Since $h > h_1 > h_2 > h_3 \dots \dots h_{n-1} > h_n$ it follows that as the order n advances, the value of h_n, or of θ_n, diminishes.

III. *Limitations to Taylor's Theorem.*

(45.) Let $R(h)$ be a function of h which represents the sum of all the terms after the *first* in the expansion of the binomial function $f(x + h)$; that is, let

$$f(x + h) = f(x) + R(h),$$

and suppose h alone to be variable; then the values of $R(h)$ and its differential coefficient or derived function $R'(h)$ will be

$$R(h) = f(x + h) - f(x)$$
$$R'(h) = f'(x + h).$$

Therefore as the value of $R(h)$ vanishes when $h = 0$, if the function $f(x)$ be continuous and finite from x to $x + h$, we have by the theorem of art. (43), or the more general theorem of art. (44),

$$R(h) = h\,R'(\theta h) = hf'(x + \theta h),$$

the value of $R'(\theta h)$ being expressed by substituting θh for h in the value of $R'(h)$;

$$\therefore\ f(x + h) = f(x) + hf'(x + \theta h) \dots \dots (1),$$

which is the development made complete in two terms.

Let now $R(h)$ be a function of h which represents the sum of all the terms after the *two first* in the development of the binomial function $f(x + h)$; that is, as suggested by equation (1), let

$$f(x + h) = fx + hf'(x) + R(h),$$

and, as before, suppose h alone to be variable ; then the values of $R(h)$ and its derived functions will be

$$R(h)\ = f(x + h) - f(x) - hf'(x)$$
$$R'(h)\ = f'(x + h) - f'(x)$$
$$R''(h) = f''(x + h).$$

Therefore as the values of $R(h)$, $R'(h)$ both vanish when $h = 0$, if $f(x)$, $f'(x)$ be continuous and finite from x to $x + h$, we have by the theorem of art. (44)

$$R(h) = \frac{h^2}{1.2}R''(\theta h) = \frac{h^2}{1.2}f''(x + \theta h)\ ;$$

$$\therefore\ f(x + h) = f(x) + hf'(x) + \frac{h^2}{1.2}f''(x + \theta h) \dots \dots (2),$$

which is the development when made complete in three terms.

Again, let $R(h)$ represent the sum of all the terms succeeding the *three first* in the development of $f(x + h)$; that is, as suggested by equation (2), let

$$f(x + h) = f(x) + hf'(x) + \frac{h^2}{1.2}f''(x) + R(h)\ ;$$

then the values of $R(h)$ and its derived functions will be

$$R(h) = f(x + h) - f(x) - hf'(x) - \frac{h^2}{1.2}f''(x)$$

$$R'(h) = f'(x + h) - f'(x) - hf''(x)$$

$$R''(h) = f''(x + h) - f''(x)$$

$$R'''(h) = f'''(x + h).$$

Hence, as the values of $R(h)$, $R'(h)$, $R''(h)$ severally vanish when $h = 0$, if $f(x)$, $f'(x)$, $f''(x)$ be continuous and finite in value from x to $x + h$, we have by the same theorem, art. (44),

$$R(h) = \frac{h^3}{1.2.3} R'''(\theta h) = \frac{h^3}{1.2.3}f'''(x + \theta h) ;$$

$$\therefore \ f(x + h) = f(x) + hf'(x) + \frac{h^2}{1.2}f''(x)$$

$$+ \frac{h^3}{1.2.3}f'''(x + \theta h) \ldots \ldots (3),$$

which is the development completed in four terms.

In like manner, so long as the functions are continuous and finite in value, may the binomial function $f(x + h)$ be completely exhibited in any number of terms. Thus, let $R(h)$ be a function of h which expresses the exact residue of the development after the *first n terms*, so that

$$f(x + h) = f(x) + \frac{h}{1}f'(x) + \frac{h^2}{1.2}f''(x) + \frac{h^3}{1.2.3}f'''(x)$$

$$\ldots \ldots \ldots + \frac{h^{n-1}}{1.2.3 \ldots n - 1}f^{(n-1)}(x) + R(h).$$

Then the values of $R(h)$ and its derived functions will be

$$R(h) = f(x + h) - f(x) - \frac{h}{1}f'(x) - \frac{h^2}{1.2}f''(x) - \frac{h^3}{1.2.3}f'''(x)$$

$$\ldots \ldots \ldots - \frac{h^{n-1}}{1.2.3 \ldots n-1}f^{(n-1)}(x)$$

$$R'(h) = f'(x+h) - f'(x) - \frac{h}{1}f''(x) - \frac{h^2}{1.2}f'''(x) \ldots \ldots$$

$$- \frac{h^{n-2}}{1.2.3 \ldots n-2}f^{(n-1)}(x)$$

$$R''(h) = f''(x+h) - f''(x) - \frac{h}{1}f'''(x) \ldots \ldots$$

$$- \frac{h^{n-3}}{1.2.3 \ldots n-3}f^{(n-1)}(x)$$

&c. &c. &c.

$$R^{(n-2)}(h) = f^{(n-2)}(x+h) - f^{(n-2)}(x) - \frac{h}{1}f^{(n-1)}(x)$$

$$R^{(n-1)}(h) = f^{(n-1)}(x+h) - f^{(n-1)}(x)$$

$$R^{(n)}(h) = f^{(n)}(x+h).$$

Therefore, when h vanishes,

$$R(0) = 0, \ R'(0) = 0, \ R''(0) = 0, \ldots \ldots R^{(n-1)}(0) = 0 ;$$

and hence if $f(x)$, $f'(x)$, $f''(x) \ldots \ldots f^{(n)}(x)$ are severally continuous and finite in value from x to $x+h$, the function $R(h)$ fulfils the conditions of the theorem of art. (44), which gives

$$R(h) = \frac{h^n}{1.2.3 \ldots n} R^{(n)}(\theta h) = \frac{h^n}{1.2.3 \ldots n}f^{(n)}(x+\theta h).$$

The development in Taylor's series, when made complete in $n+1$ terms is therefore

$$f(x+h) = f(x) + \frac{h}{1}f'(x) + \frac{h^2}{1.2}f''(x) + \frac{h^3}{1.2.3}f'''(x) \ldots \ldots$$

$$+ \frac{h^{n-1}}{1.2 \ldots n-1}f^{(n-1)}(x) + \frac{h^n}{1.2 \ldots n}f^{(n)}(x+\theta h) \ldots \ldots (n),$$

where θ is some positive numerical quantity, the value of which is undetermined further than that it is contained between the limits of *zero* and *unity*. We are hereby enabled to affix corresponding limits to the completion of Taylor's series after any number of terms; but it must be remembered, art. (41), that the value of $f^{(n)}(x+\theta h)$, though functionally intermediate, is not necessarily contained arithmetically between $f^{(n)}(x)$ and $f^{(n)}(x+h)$. Let V and v denote the greatest and least values of $f^{(n)}(x)$ which occur from x to $x + h$, then we conclude that, by stopping at the nth term, the final correction, to make the value of the development *exact*, will always be comprised between $\dfrac{h^n}{1.2\ldots n}\,\mathrm{V}$ and $\dfrac{h^n}{1.2\ldots n}\,v.$

This formula is Lagrange's limitation to Taylor's theorem, and it should be remembered that the conditions on which it depends are, that the $n + 1$ functions $f(x)$, $f'(x)$, $f''(x)$, $f'''(x)\ldots\ldots f^{(n)}(x)$ must be severally continuous and finite in value between the limits x and $x + h$. It is not affected by any of the subsequent functions $f^{(n+1)}(x)$, $f^{(n+2)}(x)$, &c. becoming discontinuous or infinite, and it is true when stopped at any number of terms, provided only that the functions are so far continuous and finite. Thus we may have

$$f(x + h) = f(x) + \frac{h}{1}f'(x + \theta_1 h)$$

$$= f(x) + \frac{h}{1}f'(x) + \frac{h^2}{1.2}f''(x + \theta_2 h)$$

$$= f(x) + \frac{h}{1}f'(x) + \frac{h^2}{1.2}f''(x) + \frac{h^3}{1.2.3}f'''(x + \theta_3 h),$$

&c.   &c.   &c.

which equations admit of being made exact by values of θ_1, θ_2, θ_3, &c., each less than unity, so that $x + \theta h$ is in every case comprised between the limits x and $x + h$. By equating each of these values of $f(x + h)$ with the next, we deduce the following relations,

$$f'(x + \theta_1 h) = f'(x) + \frac{h}{2}f''(x + \theta_2 h),$$

$$f''(x + \theta_2 h) = f''(x) + \frac{h}{3}f'''(x + \theta_3 h),$$

&c.   &c.   &c.

$$f^{(n-1)}(x + \theta_{n-1} h) = f^{(n-1)}(x) + \frac{h}{n}f^{(n)}(x + \theta_n h);$$

and from these we infer that, when h is small,

$$\theta_1 = \tfrac{1}{2}, \quad \theta_2 = \tfrac{1}{3}, \quad \theta_3 = \tfrac{1}{4}, \ldots .\theta_n = \tfrac{1}{n+1},$$

and they will seldom in any case differ much from these values.

(46.) By making $x = 0$ in the formula (n), Taylor's theorem with limits becomes

$$f(h) = f(0) + \frac{h}{1}f'(0) + \frac{h^2}{1.2}f''(0) + \frac{h^3}{1.2.3}f'''(0) \cdots$$

$$+ \frac{h^n}{1.2\ldots n}f^{(n)}(\theta h) ;$$

or, substituting x for h,

$$f(x) = f(0) + \frac{x}{1}f'(0) + \frac{x^2}{1.2}f''(0) + \frac{x^3}{1.2.3}f'''(0) \cdots$$

$$+ \frac{x^n}{1.2\ldots n}f^{(n)}(\theta x),$$

and this equation, which is necessarily exact for some value of θ less than unity, is the corresponding limitation of the theorem of Maclaurin or Stirling. The conditions essential to this theorem are, that the functions $f(x)$, $f'(x)$, $f''(x) \ldots$ $f^{(n)}(x)$ should be continuous and finite in value from 0 to x.

This theorem may also be put under the form

$$u_x = u_0 + \frac{x}{1}\left(\frac{du}{dx}\right)_0 + \frac{x^2}{1.2}\left(\frac{d^2u}{dx^2}\right)_0 + \frac{x^3}{1.2.3}\left(\frac{d^3u}{dx^3}\right)_0 \cdots$$

$$+ \frac{x^n}{1.2\ldots n}\left(\frac{d^nu}{dx^n}\right)_{\theta x}$$

IV. *Functions of Two or more Variables.*

(47.) Let $u = \mathrm{F}(x, y)$ be a function of two variables, and let it be required to expand $\mathrm{F}(x + h, y + k)$ in powers of h and k. Take $k = ah$ and put

$$\mathrm{U} = \mathrm{F}(x + h, y + k) = \mathrm{F}(x + h, y + ah).$$

Then, by supposing h alone to vary, U may be considered as a function of one variable h, and expanded in powers of h by Stirling's theorem, art. (46). When h becomes $h + dh$, the function U becomes $\mathrm{F}(x + h + dh, y + ah + adh)$, and this form is identically the same as if we had supposed x to become $x + dh$ and y to become $y + a\,dh$. Therefore, substituting dh for dx and $a\,dh$ for dy, in the formula

$$d\mathrm{U} = \left(\frac{d\mathrm{U}}{dx}\right)dx + \left(\frac{d\mathrm{U}}{dy}\right)dy,$$

we find the differential of $\mathrm{U} = \mathrm{F}(x + h,\ y + ah)$, with respect to h, to be

$$d\mathrm{U} = \left(\frac{d\mathrm{U}}{dx}\right)dh + a\left(\frac{d\mathrm{U}}{dy}\right)dh$$

$$\therefore\ \frac{d\mathrm{U}}{dh} = \left(\frac{d\mathrm{U}}{dx}\right) + a\left(\frac{d\mathrm{U}}{dy}\right)\dots\dots (1).$$

As this value of $\dfrac{d\mathrm{U}}{dh}$ must be a function of $x + h,\ y + ah$, it may evidently be again differentiated by applying to it the same formula (1). Thus

$$\frac{d}{dh}\frac{d\mathrm{U}}{dh} = \left(\frac{d}{dx}\frac{d\mathrm{U}}{dh}\right) + a\left(\frac{d}{dy}\frac{d\mathrm{U}}{dh}\right);$$

that is, operating on the preceding value of $\dfrac{d\mathrm{U}}{dh}$ as indicated on the right hand of this equation,

$$\dots\ \frac{d^2\mathrm{U}}{dh^2} = \left(\frac{d^2\mathrm{U}}{dx^2}\right) + 2a\left(\frac{d^2\mathrm{U}}{dx\,dy}\right) + a^2\left(\frac{d^2\mathrm{U}}{dy^2}\right)\dots (2).$$

In the same way, treating this as another function of $x + h$, $y + ah$, and again employing the formula (1), the process may be carried to any order of differentiation; and we shall obtain generally

$$\frac{d^n\mathrm{U}}{dh^n} = \left(\frac{d^n\mathrm{U}}{dx^n}\right) + na\left(\frac{d^n\mathrm{U}}{dx^{n-1}dy}\right) + \frac{n(n-1)}{2}a^2\left(\frac{d^n\mathrm{U}}{dx^{n-2}dy^2}\right)$$

$$\dots\dots + a^n\left(\frac{d^n\mathrm{U}}{dy^n}\right)\dots\dots (n),$$

in which the numerical coefficients are those of the expansion of $(1 + x)^n$.

Now, by Stirling's theorem with limits, art. (46), we have

$$U = U_0 + \frac{h}{1}\left(\frac{dU}{dh}\right)_0 + \frac{h^2}{1.2}\left(\frac{d^2U}{dh^2}\right)_0 \cdots\cdots\cdots$$

$$+ \frac{h^n}{1.2\ldots n}\left(\frac{d^nU}{dh^n}\right)_{\theta h}$$

in which expansion the function U and its differential coefficients are the values when $h = 0$, excepting the last, in which h takes the value θh. But when $h = 0$, functions of $x + h$, $y + ah$ become corresponding functions of x, y, and U, and its differential coefficients with respect to x and y become the same as if the function u had been employed; also when h becomes θh, functions of $x + h$, $y + ah$ become corresponding functions of $x + \theta h$, $y + \theta ah$. Hence substituting the values according to the preceding expressions (1), (2), $\ldots\ldots(n)$, and observing these transformations, we have for U the following development:

$$U = F(x + h, y + ah) =$$

$$u + \frac{h}{1}\left\{\left(\frac{du}{dx}\right) + a\left(\frac{du}{dy}\right)\right\}$$

$$+ \frac{h^2}{1.2}\left\{\left(\frac{d^2u}{dx^2}\right) + 2a\left(\frac{d^2u}{dx\,dy}\right) + a^2\left(\frac{d^2u}{dy^2}\right)\right\}$$

$$\cdot\ \cdot\ \cdot\ \cdot\ \cdot\ \cdot\ \cdot\ \cdot\ \cdot\ \cdot\ \cdot\ \cdot\ \cdot\ \cdot\ \cdot$$

$$+ \frac{h^n}{1.2\ldots n}\left\{\left(\frac{d^nu}{dx^n}\right) + na\left(\frac{d^nu}{dx^{n-1}dy}\right)\right.$$

$$+ \frac{n(n-1)}{2}a^2\left(\frac{d^nu}{dx^{n-2}dy^2}\right)$$

$$\left.\cdots\cdots + a^n\left(\frac{d^nu}{dy^n}\right)\right\}_{\substack{x+\theta h\\y+\theta ah}}$$

the value of the term exhibited in the last three lines being taken when x and y become $x + \theta h$, $y + \theta ah$, where $\theta < 1$.

D

By substituting k in place of its value ah, the formula becomes

$$U = F(x + h, y + k) =$$

$$u + h\left(\frac{du}{dx}\right) + k\left(\frac{du}{dy}\right)$$

$$+ \frac{1}{1.2}\left\{ h^2\left(\frac{d^2u}{dx^2}\right) + 2hk\left(\frac{d^2u}{dx\,dy}\right) + k^2\left(\frac{d^2u}{dy^2}\right) \right\}$$

$$\cdots\cdots\cdots\cdots\cdots\cdots\cdots\cdots$$

$$+ \frac{1}{1.2\ldots.n}\left\{ h^n\left(\frac{d^nu}{dx^n}\right) + nh^{n-1}k\left(\frac{d^nu}{dx^{n-1}dy}\right) \right.$$

$$+ \frac{n(n-1)}{2}h^{n-2}k^2\left(\frac{d^nu}{dx^{n-2}dy^2}\right)$$

$$\left. \cdots\cdots + k^n\left(\frac{d^nu}{dy^n}\right) \right\}_{\substack{x+\theta h \\ y+\theta k}}$$

(48.) In the formula just determined make $x = 0$, $y = 0$, and afterwards change h into x and k into y; then

$$u = F(x, y) = u_0 + x\left(\frac{du}{dx}\right)_0 + y\left(\frac{du}{dy}\right)_0$$

$$+ \frac{1}{1.2}\left\{ x^2\left(\frac{d^2u}{dx^2}\right)_0 + 2xy\left(\frac{d^2u}{dx\,dy}\right)_0 + y^2\left(\frac{d^2u}{dy^2}\right)_0 \right\}$$

$$\cdots\cdots\cdots\cdots\cdots\cdots$$

$$+ \frac{1}{1.2\ldots.n}\left\{ x^n\left(\frac{d^nu}{dx^n}\right) + nx^{n-1}y\left(\frac{d^nu}{dx^{n-1}dy}\right) \right.$$

$$\left. \cdots\cdots + y^n\left(\frac{d^nu}{dy^n}\right) \right\}_{\substack{\theta x \\ \theta y}}$$

where we have to make x, y each $= 0$ in the several functions, except in the term which occupies the last two lines, where they are to be replaced by θx, θy, θ being < 1.

Note.—It may here be remarked with respect to expansions

generally, that if the nth or limiting term decreases without limit as n increases without limit, the development may be then continued without introducing any limiting term.

(49.) If in Taylor's theorem we make $h = dx$, it becomes

$$f(x + dx) = f(x) + \frac{df(x)}{1} + \frac{d^2f(x)}{1.2} + \frac{d^3f(x)}{1.2.3} + \&c. \, ;$$

that is, if $u = f(x)$,

$$U = f(x + dx) = u + \frac{du}{1} + \frac{d^2u}{1.2} + \frac{d^3u}{1.2.3} + \&c.$$

This formula represents in a simple form the most general theory of expansion, and may be extended to the expansion of a function of any number of variables, under the following general enunciation :

* Let $u = f(x, y, z, \&c.)$ be a function of any number of variables, and let $\delta x, \delta y, \delta z, \&c.$ denote arbitrary increments of the respective variables.

Suppose the function

$$U = f(x + \delta x, y + \delta y, z + \delta z, \&c.)$$

to be partly expanded, and denote by δu the terms which involve the first order of the increments $\delta x, \delta y, \delta z, \&c.$

Then $x + \delta x, y + \delta y, z + \delta z, \&c.$ being substituted for $x, y, z, \&c.$ in the value of δu and the result again partly expanded, denote by $\delta^2 u$ the terms which involve the second order of the increments.

And again, the same substitutions being made in $\delta^2 u$, and the result expanded, denote by $\delta^3 u$ the terms which involve the third order of the increments, &c., &c.

Then will

$$U = u + \frac{\delta u}{1} + \frac{\delta^2 u}{1.2} + \frac{\delta^3 u}{1.2.3} + \&c. \, ;$$

and the values of $\delta u, \delta^2 u, \delta^3 u, \&c.$ may be determined by successively differentiating the function $u = f(x, y, z, \&c.)$ on

* This theorem was first announced by the author in the Appendix to the 'Gentleman's Diary' for the year 1835.

the supposition that dx, dy, dz, &c. do not change, only writing δx, δy, δz, &c. in place of dx, dy, dz, &c. ; also the series may be stopped at pleasure by substituting $x + \theta \delta x$, $y + \theta \delta y$, $z + \theta \delta z$, &c. for x, y, z, &c. in the last term, θ being < 1.

By making x, y, z, &c. severally $= 0$, and writing x, y, z, &c. in place of δx, δy, δz, &c., the result will be the expansion of the function $u = f(x, y, z,$ &c.) in powers of the variables.

The preceding developments may all be deduced from this general theorem.

Examples.

1. Expand $f(x + h) = (x + h)^n$ by Taylor's theorem.
Since $f(x) = x^n$, we have by successive differentiation

$$f'(x) = n x^{n-1}, \qquad f''(x) = n(n-1) x^{n-2},$$
$$f'''(x) = n(n-1)(n-2) x^{n-3}, \&c.$$

Hence, by the theorem, art. (39),

$$(x + h)^n = x^n + \frac{n}{1} x^{n-1} h + \frac{n(n-1)}{1.2} x^{n-2} h^2$$
$$+ \frac{n(n-1)(n-2)}{1.2.3} x^{n-3} h^3 + \&c.,$$

which is the formula of the binomial theorem.

2. Expand $\log(x + h)$.
Here $f(x) = \log x$, and by differentiation

$$f'(x) = x^{-1}, \qquad f''(x) = -1 . x^{-2}, \qquad f'''(x) = 1.2.x^{-3}, \&c.$$

Therefore, by the theorem,

$$f(x + h) = \log(x + h) = \log(x) + \frac{h}{x} - \frac{h^2}{2 x^2} + \frac{h^3}{3 x^3} - \&c.$$

which is divergent and inapplicable when $x < h$.

If we employ the theorem with the limitations, art. (45), we shall obtain

$$\log(x + h) = \log(x) + \frac{h}{x + \theta h}$$

$$= \log (x) + \frac{h}{x} - \frac{h^2}{2(x + \theta h)^2},$$

which expressions will be strictly accurate with values of θ between the limits of zero and unity. Let $x = 1$, then

$$\log (1 + h) = \frac{h}{1 + \theta h} = h - \frac{h^2}{2(1 + \theta h)^2}.$$

By the first of these expressions it follows that the value of $\log (1 + h)$ is comprised between $\frac{h}{1}$ and $\frac{h}{1 + h}$; and by the second the same value is comprised between the narrower limits $h - \frac{h^2}{2}$ and $h - \frac{h^2}{2(1 + h)^2}$.

3. Expand the function $u = \sin x$ in powers of x by Maclaurin's theorem.

By differentiation,

$$\frac{du}{dx} = \cos x, \qquad \frac{d^2 u}{dx^2} = - \sin x, \qquad \frac{d^3 u}{dx^3} = - \cos x,$$

$$\frac{d^4 u}{dx^4} = \sin x, \qquad \frac{d^5 u}{dx^5} = \cos x, \qquad \text{\&c.}$$

which, when $x = 0$, respectively become 1, 0, $- 1$, 0, 1, &c. Therefore by the theorem, art. (40),

$$\sin x = x - \frac{x^3}{1.2.3} + \frac{x^5}{1.2.3.4.5} - \text{\&c.}$$

Or, by the theorem with limitations, art. (46),

$$\sin x = x \cos \theta x = x - \frac{x^2}{1.2} \sin \theta_{,} x; \text{ where } \theta_{,} < \theta < 1,$$

and which may be similarly expressed in any required number of terms.

4. Expand $u = \cos x$, in powers of x.

Here $\frac{du}{dx} = - \sin x, \qquad \frac{d^2 u}{dx^2} = - \cos x,$

$$\frac{d^3 u}{dx^3} = \sin x, \qquad \frac{d^4 u}{dx^4} = \cos x, \text{ \&c.}$$

which, when $x = 0$, become $0, -1, 0, 1,$ &c.;

$$\therefore \cos x = 1 - \frac{x^2}{1.2} + \frac{x^4}{1.2.3.4} - \text{&c.}$$

Or, with the limitations,

$$\cos x = 1 - x \sin \theta x = 1 - \frac{x^2}{1.2} \cos \theta_{,}x = \text{&c.}$$

5. Expand $u = e^x = \log^{-1}x$ in powers of x.

By art. (26) we have $\dfrac{du}{dx} = e^x$, $\dfrac{d^2u}{dx^2} = e^x$, &c., which, when $x = 0$, severally become equal to unity.

$$\therefore e^x = 1 + \frac{x}{1} + \frac{x^2}{1.2} + \frac{x^3}{1.2.3} + \text{&c.}$$

Also, with the limitations,

$$e^x = 1 + \frac{x}{1} e^{\theta x} = 1 + \frac{x}{1} + \frac{x^3}{1.2} e^{\theta_{,}x} = \text{&c.}$$

6. Let $u = x\,y\,z$, and expand

$$U = (x + \delta x)\,(y + \delta y)\,(z + \delta z)$$

by the general theorem of art. (49).

By operating upon $u = x\,y\,z$ with the symbol δ in a manner analogous to successive differentiation, and supposing δx, δy, δz to be invariable, we have

$$u = x\,y\,z$$
$$\delta u = y\,z\,\delta x + z\,x\,\delta y + x\,y\,\delta z$$
$$\delta^2 u = 2\,x\,\delta y\,\delta z + 2\,y\,\delta z\,\delta x + 2\,z\,\delta x\,\delta y$$
$$\delta^3 u = 6\,\delta x\,\delta y\,\delta z,$$

which substituted in the formula

$$U = u + \frac{\delta u}{1} + \frac{\delta^2 u}{1.2} + \frac{\delta^3 u}{1.2.3} + \text{&c.}$$

we obtain

$$(x + \delta x)\,(y + \delta y)\,(z + \delta z) = x\,y\,z + (y\,z\,\delta x + z\,x\,\delta y + x\,y\,\delta z)$$
$$+ (x\,\delta y\,\delta z + y\,\delta z\,\delta x + z\,\delta x\,\delta y)$$
$$+ \delta x\,\delta y\,\delta z,$$

which may be verified by multiplication.

(50.) In the series for e^x, example 5, replace x by $x\sqrt{-1}$; then

$$e^{x\sqrt{-1}} = 1 + x\sqrt{-1} - \frac{x^2}{1.2} - \frac{x^3}{1.2.3}\sqrt{-1} + \frac{x^4}{1.2.3.4} + \&c.$$

$$= 1 - \frac{x^2}{1.2} + \frac{x^4}{1.2.3.4} - \&c.$$

$$+ \left(x - \frac{x^3}{1.2.3} + \&c.\right)\sqrt{-1};$$

that is, examples 3 and 4,

$$e^{x\sqrt{-1}} = \cos x + \sqrt{-1}\sin x \dots \dots \dots (1).$$

In this equation replace x by $-x$, and we have also

$$e^{-x\sqrt{-1}} = \cos x - \sqrt{-1}\sin x \dots \dots \dots (2);$$

$$\therefore \quad \cos x = \frac{e^{x\sqrt{-1}} + e^{-x\sqrt{-1}}}{2},$$

$$\left. \sin x = \frac{e^{x\sqrt{-1}} - e^{-x\sqrt{-1}}}{2\sqrt{-1}} \right\} \dots \dots (3),$$

which are Euler's formulæ.

Again, replacing x by mx in (1) and (2),

$$e^{\pm mx\sqrt{-1}} = \cos mx \pm \sqrt{-1}\sin mx.$$

Hence, as $e^{\pm mx\sqrt{-1}} = (e^{\pm x\sqrt{-1}})^m$, we have

$$\cos mx \pm \sqrt{-1}\sin mx = (\cos x \pm \sqrt{-1}\sin x)^m \dots \dots (4),$$

which is De Moivre's formula and is true for all integral values of m. When expanded by the binomial theorem, by equating separately the real and the unreal portions, we may obtain from it the trigonometrical values of $\cos mx$ and $\sin mx$ in powers of $\cos x$, $\sin x$.

In (4) replace x by $x + 2r\pi$, r denoting any integral number; then

$$(\cos x \pm \sqrt{-1}\sin x)^m =$$

$$\cos(mx + 2rm\pi) \pm \sqrt{-1}\sin(mx + 2rm\pi) \dots (5),$$

which is the complete form of equation (4) and is now true for all values of m, whether integral, fractional, real or unreal; and both sides will now always contain the same number of identical values.*

From the preceding values of $\cos x$, $\sin x$, equations (3), it is evident that all the trigonometrical functions of x may be expressed in algebraical functions of the exponentials $e^{x\sqrt{-1}}$ and $e^{-x\sqrt{-1}}$.

CHAPTER V.

INDETERMINATE FORMS.

(51.) When a function for a particular value of the variable assumes any one of the forms

$$\frac{0}{0}, \frac{\infty}{\infty}, 0 \times \infty, \infty - \infty; \ 0^0, \infty^0 \text{ or } 1^{\pm \infty},$$

the function, absolutely considered under this singular form, becomes then essentially indeterminate and admits of having any value whatever assigned to it. But if the proposed function represent a quantity which varies continuously so that the function up to the particular value of the variable is subject to a condition of continuity, its value will evidently be determinable in a manner analogous to that by which we obtained the differential coefficient of a function in art. (6).

I. *Functions in the Form of Fractions.*

(52.) Let $u = \dfrac{f(x)}{F(x)}$ be a function of x which becomes $\dfrac{0}{0}$ when $x = a$. It is evident that this will arise from the incorporation of certain vanishing factors in both numerator and

---

* An investigation of the general theory of exponential and imaginary quantities arising out of this last equation is given by the author in the Appendix to the ' Gentleman's Diary ' for 1837.

denominator. Suppose the resolution of these factors to give

$$\frac{f(x)}{F(x)} = \frac{(x-a)^m P}{(x-a)^n Q},$$

where P and Q are of finite value when $x = a$. Then by division we should have

$$\frac{f(x)}{F(x)} = (x-a)^{m-n} \frac{P}{Q};$$

and when $x = a$, this would obviously give for the required value,

$$0 \text{ if } m > n; \quad \frac{P}{Q} \text{ if } m = n, \text{ or } \infty \text{ if } m < n.$$

The elimination of the vanishing factors will in most cases be facilitated by substituting $a + h$ for x, so that $x - a = h$. The form of u will then be a function of h which becomes $\frac{0}{0}$ when $h = 0$. By expanding, if necessary, the numerator and denominator of this function in ascending powers of h, and dividing by the power of h which is common to them both, and afterwards making $h = 0$, the result will be the required continuous value of the proposed vanishing fraction when $x = a$.

(53.) The continuous value of the vanishing fraction may be otherwise determined by ascertaining in a different manner an expression of its value in a continuous form for values of x contiguous to $x = a$. Thus when x takes the value $a + h$, we have by Taylor's theorem, art. (45), observing that $f(a) = 0$, $F(a) = 0$,

$$\frac{f(a+h)}{F(a+h)} = \frac{f(a) + \frac{h}{1} f'(a+\theta h)}{F(a) + \frac{h}{1} F'(a+\theta h)} = \frac{f'(a+\theta h)}{F'(a+\theta h)}.$$

This equation is necessarily strictly true when h is of any value, however small, positive or negative, and if $f'(a)$,

$F'(a)$ do not both vanish or become infinite, the fraction on the right hand will be continuous in form when h vanishes; therefore, making $h = 0$, we obtain, for the continuous value,

$$\frac{f(a)}{F(a)} = \frac{f'(a)}{F'(a)} \quad \ldots \ldots (1).$$

But if $f'(a)$, $F'(a)$ both vanish, by extending Taylor's series to another term, we shall have

$$\frac{f(a+h)}{F(a+h)} = \frac{f(a) + \frac{h}{1} f'(a) + \frac{h^2}{1.2} f''(a + \theta h)}{F(a) + \frac{h}{1} F'(a) + \frac{h^2}{1.2} F''(a + \theta h)}$$

$$= \frac{f''(a + \theta h)}{F''(a + \theta h)}.$$

Hence, if $f''(a)$, $F''(a)$ do not both vanish or become infinite, we obtain, by making $h = 0$,

$$\frac{f(a)}{F(a)} = \frac{f''(a)}{F''(a)} \quad \ldots \ldots \ldots (2).$$

By proceeding in this way, we similarly find that if the numerator and denominator with their first $n-1$ differential coefficients, viz. $f(x), f'(x), f''(x) \ldots f^{(n-1)}(x)$, and $F(x)$, $F'(x)$, $F''(x) \ldots \ldots F^{(n-1)}(x)$ severally vanish when $x = a$, and the nth differential coefficients $f^{(n)}(x)$, $F^{(n)}(x)$ do not both vanish or become infinite, then the continuous value of the fraction will be

$$\frac{f(a)}{F(a)} = \frac{f^{(n)}(a)}{F^{(n)}(a)} \quad \ldots \ldots (n).$$

(54.) Suppose the numerator and denominator of the function $\dfrac{f(x)}{F(x)}$ to be both of them infinite in value when $x = a$, so that it becomes of the form $\dfrac{\infty}{\infty}$. Then by expressing the function by the reciprocals, thus,

$$\frac{f(a)}{F(a)} = \frac{\dfrac{1}{F(a)}}{\dfrac{1}{f(a)}},$$

it will become of the form $\dfrac{0}{0}$. Therefore by equation (1) we get, by differentiating the numerator and denominator,

$$\frac{f(a)}{F(a)} = \frac{-\dfrac{F'(a)}{\{F(a)\}^2}}{-\dfrac{f'(a)}{\{f(a)\}^2}} = \left\{\frac{f(a)}{F(a)}\right\}^2 \frac{F'(a)}{f'(a)},$$

which gives

$$\frac{f(a)}{F(a)} = \frac{f'(a)}{F'(a)}.$$

This being the same as the equation (1) before obtained, we conclude that the mode of operating in this case is identical with that already indicated when the function is of the form $\dfrac{0}{0}$.

Thus, if after $n-1$ differentiations the fractions $\dfrac{f'(a)}{F'(a)}$, $\dfrac{f''(a)}{F''(a)}$, $\dfrac{f'''(a)}{F'''(a)}$, $\cdots\cdots$ $\dfrac{f^{(n-1)}(a)}{F^{(n-1)}(a)}$ severally become of the form $\dfrac{\infty}{\infty}$ or $\dfrac{0}{0}$, and if $\dfrac{f^{(n)}(a)}{F^{(n)}(a)}$ does not become of either of those forms; then, according to equation (n),

$$\frac{f(a)}{F(a)} = \frac{f^{(n)}(a)}{F^{(n)}(a)}.$$

(55.) We have therefore the following rule for determining the continuous value of a fraction which for a particular value of the variable becomes of the form $\dfrac{0}{0}$ or $\dfrac{\infty}{\infty}$:—Divide the differential coefficient of the numerator by the differential coefficient of the denominator for a new fraction, in which substitute the given value of the variable. Should this latter fraction still assume the form $\dfrac{0}{0}$ or $\dfrac{\infty}{\infty}$, the same process may be suc-

cessively repeated until one or both of the numerator and
denominator ceases to vanish or become infinite in value.

Example 1.—When $x = 0$, find the continuous value of
$$\frac{1 - \cos x}{\sin^2 x} = \frac{0}{0}.$$

Here $fx = 1 - \cos x$, $F(x) = \sin^2 x$; and by differentiation,
$$\frac{f'(x)}{F'(x)} = \frac{\sin x}{2 \sin x \cos x} = \frac{1}{2 \cos x},$$
which, when $x = 0$, gives $\frac{1}{2}$ for the required value.

Example 2.—When $x = 0$, required the value of $\dfrac{\log \sin x}{\log \sin 2x}$
$$= \frac{-\infty}{-\infty}.$$

Since $f(x) = \log \sin x$, $F(x) = \log \sin 2x$, we have
$$f'x = \frac{\cos x}{\sin x}, \quad F'(x) = \frac{2 \cos 2x}{\sin 2x};$$
$$\therefore \frac{f'(x)}{F'(x)} = \frac{\cos x}{2 \cos 2x} \cdot \frac{\sin 2x}{\sin x}.$$

When $x = 0$, the first factor of this expression is determi-
nate and is $\dfrac{\cos x}{2 \cos 2x} = \frac{1}{2}$; but the other factor $\dfrac{\sin 2x}{\sin x}$ still

maintains the indeterminate form $\dfrac{0}{0}$, and its numerator and
denominator must therefore be again differentiated, giving
$\dfrac{2 \cos 2x}{\cos x} = 2.$ The value of the proposed expression is
therefore $\frac{1}{2} \times 2 = 1$.

Example 3.—When $x = \infty$, determine the continuous value
of $\dfrac{e^x}{x^m} = \dfrac{\infty}{\infty}$, the exponent m being a finite integer.

Here we have $\dfrac{f(x)}{F(x)} = \dfrac{e^x}{x^m} = \dfrac{\infty}{\infty}$, when $x = \infty$,
$$\frac{f'(x)}{F'(x)} = \frac{e^x}{m x^{m-1}} = \frac{\infty}{\infty}, \text{ when } x = \infty,$$
&c.   &c.   &c.

$$\frac{f^{(m)}(x)}{F^{(m)}(x)} = \frac{e^x}{1.2.3 \ldots m} = \infty, \text{ when } x = \infty$$

The sought value is therefore infinite.

4. When $x = 1$, then $\dfrac{1 - x^m}{1 - x^n} = \dfrac{0}{0} = \dfrac{m}{n}$.

5. When $x = a$, then $\dfrac{e^x - e^a}{x - a} = \dfrac{0}{0} = e^a$.

6. When $x = 0$, then $\dfrac{a^x - b^x}{x} = \dfrac{0}{0} = \log \dfrac{a}{b}$.

7. When $x = 0$, then $\dfrac{e^x - e^{-x}}{\sin x} = \dfrac{0}{0} = 2$.

8. When $x = 0$, then $\dfrac{x - \sin x}{x^3} = \dfrac{0}{0} = \dfrac{1}{2.3}$.

9. When $x = 0$, then $\dfrac{\tan x - \sin x}{x - \sin x} = \dfrac{0}{0} = 3$.

10. When $x = 1$, then $\dfrac{x^x - x}{1 + \log x - x} = \dfrac{0}{0} = -2$.

11. When $x = 0$, then $\dfrac{\log \cot x}{\log x} = \dfrac{\infty}{-\infty} = -1$.

12. When $x = 0$, then $\dfrac{\cos ax - \cos \beta x}{\cos ax - \cos bx} = \dfrac{0}{0} = \dfrac{a^2 - \beta^2}{a^2 - b^2}$.

II. *Functions in the Form of Products.*

(56.) Again, if $F(x) f(x)$ be a function of x which, when $x = a$, becomes $0 \times \infty$, it may be differently expressed, as follows:

$$F(x) f(x) = \frac{F(x)}{\dfrac{1}{f(x)}} = \frac{f(x)}{\dfrac{1}{F(x)}}.$$

Since, when $x = a$, $F(x) = 0$, $f(x) = \infty$, the former of these will assume the form $\dfrac{0}{0}$, and the latter will assume the form $\frac{\infty}{\infty}$, and either of them may be evaluated by art. (55).

Also, if $F(x) - f(x)$ be a function of x which, when $x = a$, becomes of the form $\infty - \infty$, it may be expressed thus:

$$F(x) - f(x) = \frac{\dfrac{1}{f(x)} - \dfrac{1}{F(x)}}{\dfrac{1}{F(x).f(x)}},$$

which, when $x = a$, will now become $\frac{0}{0}$, and may therefore be evaluated as before.

Example 1.—When $x = \frac{\pi}{2}$, required the value of

$$\left(1 - \frac{2x}{\pi}\right) \tan x = 0 \times \infty.$$

In this example we have

$$\left(1 - \frac{2x}{\pi}\right) \tan x = \frac{1 - \dfrac{2x}{\pi}}{\cot x}.$$

When $x = \frac{\pi}{2}$, this expression assumes the form $\frac{0}{0}$, and its value is hence found to be

$$\frac{1 - \dfrac{2x}{\pi}}{\cot x} = \frac{-\dfrac{2}{\pi}}{-\operatorname{cosec}^2 x} = \frac{2}{\pi}.$$

Example 2.—When $x = 1$, find the value of $\frac{x}{\log x} - \frac{1}{\log x}$ $= \infty - \infty$.

Here $\dfrac{x}{\log x} - \dfrac{1}{\log x} = \dfrac{x-1}{\log x}$,

which, when $x = 1$, takes the form $\frac{0}{0}$, and its value is therefore found to be

3. When $x = 1$, then $\dfrac{2}{1 - x^2} - \dfrac{1}{1 - x} = \infty - \infty = \frac{1}{2}$.

4. When $x = \infty$, then $e^{-x}\log x = 0 \times \infty = 0$.

5. When $x = 0$, then $x \log x = 0 \times -\infty = 0$.

6. When $x = 1$, then $\dfrac{x}{x-1} - \dfrac{1}{\log x} = \infty - \infty = \frac{1}{2}$.

7. When $x = 0$, then $\dfrac{1}{\sin^2 x} - \dfrac{1}{x^2} = \infty - \infty = \frac{1}{3}$.

8. When $x = 0$, then $\dfrac{1}{x^2} - \dfrac{1}{x \tan x} = \infty - \infty = \frac{1}{3}$.

III. *Functions in the Form of Exponentials.*

(57.) The general exponential function $u = F(x)^{f(x)}$ may for a particular value of x become one or other of the forms

$$0^0, \quad \infty^0, \quad 1^{\pm\infty}, \quad 0^{\pm\infty}, \quad \infty^{\pm\infty}.$$

Only the first three of these are indeterminate in their character: the other two are determinate, and their values are evidently

$$0^{\pm\infty} = \left\{ \begin{array}{l} 0 \\ \infty \end{array} \right. \qquad \infty^{\pm\infty} = \left\{ \begin{array}{l} \infty \\ 0. \end{array} \right.$$

Since $u = F(x)^{f(x)}$, we have

$$\log u = f(x) \log F(x) = \dfrac{\log F(x)}{\dfrac{1}{f(x)}}.$$

Therefore, referring to this expression for $\log u$,

when u is of the form $\left\{ \begin{array}{c} 0^0 \\ \infty \\ \pm\infty \end{array} \right\}$ $\log u$ is of the form $\left\{ \begin{array}{c} \dfrac{-\infty}{\infty} \\ \dfrac{\infty}{\infty} \\ \dfrac{0}{0}. \end{array} \right.$

Hence the value of $\log u$ may be determined by art. (55), and thence the corresponding value of u.

Example 1.—When $x = 0$, find the value of $x^x = 0^0$.

Here $u = x^x$, and $\log u = x \log x = \dfrac{\log x}{\dfrac{1}{x}}$.

When $x = 0$, this expression for $\log u$ takes the form $\dfrac{-\infty}{\infty}$, and hence, by differentiation, its value is found to be

$$\log u = \frac{\log x}{\dfrac{1}{x}} = \frac{\dfrac{1}{x}}{-\dfrac{1}{x^2}} = -x = 0 ; \quad \therefore u = 1.$$

2. When $x = 0$, then $x^{\sin x} = 0^0 = 1$.

3. When $x = 0$, then $(\cot x)^{\sin x} = \infty^0 = 1$.

4. When $x = \infty$, then $x^{\frac{1}{\log m x}} = \infty^0 = e$.

5. When $x = 0$, then $(1 + mx)^{\frac{1}{x}} = 1^\infty = e^m$.

6. When $x = 1$, then $x^{\frac{1}{1-x}} = 1^\infty = \dfrac{1}{e}$.

IV. *Exceptions to Taylor's Theorem.*

(58.) In art. (39) allusion has been made to the existence of certain functions, to the development of which Taylor's theorem ceases to be applicable for particular values of the variable, in consequence of the differential coefficients or derived functions becoming infinite in value.

Let $\psi(x)$ be a function of x, and suppose a given finite value a to be a root of either of the equations

$$\psi(x) = 0, \qquad \frac{1}{\psi(x)} = 0 ;$$

then it may be shown that $\psi(x)$ will be of the form

$$\psi(x) = (x - a)^\mu \phi(x) \quad \ldots \quad (1),$$

the function $\phi(x)$ not vanishing or becoming infinite when $x = a$, and therefore not involving as a factor any other power of $x - a$. Also, the exponent μ will be positive or negative according as $x = a$ causes $\psi(x)$ to become zero or infinity, or according as a is a root of $\psi(x) = 0$ or of $\dfrac{1}{\psi(x)} = 0$; and it will evidently be the limiting value of the fraction $\dfrac{\log \psi(x)}{\log(x-a)}$, which assumes the form $\dfrac{\infty}{\infty}$, when $x = a$.

(59.) Suppose a given function $f(x)$ to contain a term of the form $\psi(x)$; then, if we proceed to the derived functions,

$f'(x)$ will contain the term $(x - a)^{\mu-1}\phi(x) \cdot \mu$

$f''(x)$,,   ,, $(x - a)^{\mu-2}\phi(x) \cdot \mu(\mu-1)$

$f'''(x)$,,   ,, $(x - a)^{\mu-3}\phi(x) \cdot \mu(\mu-1)(\mu-2)$

     &c.   &c.   &c.

Consider now the following cases :

1. If μ be a positive whole number, these terms will wholly disappear after $f^\mu(x)$, and since the exponents $\mu - 1$, $\mu - 2$, $\mu - 3$, &c. are all positive, it is evident that when $x = a$ and $x - a = 0$, the original introduction of the factor $(x - a)^\mu$ cannot thus affect the finite character of the values of the derived functions. This case therefore does not form an exception to Taylor's theorem.

2. If μ be of the form $m + \dfrac{p}{q}$, a positive whole number with the addition of a finite fraction, then the exponents $\mu - 1$, $\mu - 2$, $\mu - 3$, &c. of the factor $(x - a)$ in the above terms will be positive for the first m derived functions, but will afterwards become negative. Therefore, when $x = a$, the terms will vanish from the first m derived functions and will become infinite in value in all the subsequent functions.

Hence, as regards the factor $(x - a)^{m+\frac{p}{q}}$, the derived functions will, when $x = a$, be finite up to $f^{(m)}(x)$, but $f^{(m+1)}(x)$ and all the subsequent functions will be infinite. The expansion of the proposed function by Taylor's theorem, for the particular value $x = a$, will therefore not in this case admit of being

carried to any terms beyond $\dfrac{h^m}{1.2 \ldots m} f^{(m)}(x + \theta h)$, and it
may be stopped at any previous term $\dfrac{h^n}{1.2 \ldots n} f^{(n)}(x + \theta h)$,
where $n < m$. Within these limits the accuracy of the development will not be affected by the infinite values of the higher derived functions.

3. If μ have a negative value, or a positive value less than unity, then the exponents $\mu - 1$, $\mu - 2$, $\mu - 3$, &c. will be all negative, and when $x = a$ all the derived functions will become infinite in value, so that the conditions of Taylor's theorem not being fulfilled, it will be wholly inapplicable to the development of the proposed function for the particular value $x = a$; but the application will nevertheless be true in all cases for values of x which differ from a by a finite quantity.

The cause of these singular results may be ascertained by examining the effect produced upon the form of the function proposed for development. Thus when $f(x)$ contains the term $(x - a)^\mu \phi(x)$, $f(x + h)$ will contain the corresponding term $(x + h - a)^\mu \phi(x + h)$, and, when $x = a$, this will become $h^\mu \phi(a + h)$. As $\phi(a)$ cannot $= 0$ or ∞, the expansion of this term will give a series involving powers of h beginning with h^μ: when μ is a positive integral number, no peculiarity is induced; but when μ is positive and fractional, all the powers of h will likewise be fractional, and when μ is negative, the development will contain negative powers of h to the same extent.

In these remarks, which apply equally to Stirling's theorem, the symbol μ, to observe the utmost generality, might have been considered as a function of x, and it is evident that all the peculiarities of form and result would then be determined in exactly the same way and would similarly depend upon the particular value of μ when $x = a$.

(60.) From what precedes we are led to the following general conclusions:

If when the variable x takes the finite value a, the function

$f(x)$ and its first m derived functions be finite and the $m + 1$th derived function be infinite; then all the succeeding derived functions will likewise be infinite, and Taylor's theorem with the limitations, art. (45), will be correct if not carried further than the term involving h^m. Beyond this term the theorem will be inapplicable, as indicated by the infinite values of the differential coefficients, because the further expansion of the proposed function $f(x + h)$ will consist of fractional powers of h, the first fractional exponent being contained between m and $m + 1$.

If when $x = a$ the value of the function itself be infinite, then the values of all the derived functions will likewise be infinite, and the true expansion will contain negative powers of h.

In either of these exceptional cases the definite expansion of the proposed function $f(x + h)$ for $x = a$ may be generally obtained by first substituting a in place of x and afterwards expanding the reduced result, supposing a to be variable, for which Taylor's theorem may be employed if necessary.

Example.—Let $f(x) = x^3 + (x^2 - a^2)^{\frac{3}{2}}$; then $f'(x)$ will involve $(x^2 - a^2)^{\frac{1}{2}}$, and $f''(x)$ will involve $(x^2 - a^2)^{-\frac{1}{2}}$ and become infinite when $x = a$.

Therefore the true expansion of $f(x + h)$ when $x = a$ will contain fractional powers of h commencing from an exponent between 1 and 2. To determine this expansion, we have

$$f(x + h) = (x + h)^3 + \{(x + h)^2 - a^2\}^{\frac{3}{2}}$$

$$\therefore f(a + h) = (a + h)^3 + \{(a + h)^2 - a^2\}^{\frac{3}{2}}$$

$$= (a + h)^3 + (2ah + h^2)^{\frac{3}{2}}$$

$$= (a + h)^3 + h^{\frac{3}{2}}(2a + h)^{\frac{3}{2}},$$

which may be readily expanded by the binomial theorem.

Again, suppose $\psi(x)$ to be of the form $e^{-\frac{1}{x^m}} \phi(x)$, where m

is positive and finite and $\phi(x)$ not $= 0$ or ∞ when $x = 0$.

Since $e^{\log x} = x$, or $e = x^{\frac{1}{\log x}}$, this function may be transformed into the equivalent expression $\psi(x) = x^{-\frac{1}{x^m \log x}} \phi(x)$;

$$\therefore \ a = 0, \text{ and } \mu = -\frac{1}{x^m \log x}.$$

When $x = a = 0$, the particular value of the function

$$\mu = \frac{-\dfrac{1}{x^m}}{\log x},$$ which takes the form $\frac{\infty}{\infty}$, must be determined by differentiating the numerator and denominator according to

art. (55); thus we find $\mu = \dfrac{\dfrac{m}{x^{m+1}}}{\dfrac{1}{x}} = \dfrac{m}{x^m}$. Hence, making

$x = 0$, the particular value of μ is infinite, so that if x were considered as an infinitesimal, the value of the function $\psi(x)$ would become an infinitesimal of an infinite order. Therefore the values of $\psi(x)$ and all its differential coefficients or derived functions will vanish when $x = 0$, and the expansion by Taylor's theorem will in this case not fail.

v. *Differential Coefficients of the form* $\dfrac{0}{0}$.

(61.) When two variables x and y are implicitly related by an equation

$$u = f(x, y) = 0,$$

let the partial differential coefficients with respect to x and y be

$$\left(\frac{du}{dx}\right) = \mathrm{P}, \qquad \left(\frac{du}{dy}\right) = \mathrm{Q};$$

then, the value of the differential coefficient or differential ratio $\dfrac{dy}{dx}$, art. (32), will be

$$\frac{dy}{dx} = -\frac{\mathrm{P}}{\mathrm{Q}}.$$

If values of x and y can be found which will fulfil the three equations $u = 0$, $P = 0$, $Q = 0$, we shall have, for these particular values,

$$\frac{dy}{dx} = \frac{0}{0},$$

and the determination of the continuous value in this case may be found by successively differentiating the numerator and denominator of the fraction, as in art. (55), with this difference that the result will lead to an equation involving $\frac{dy}{dx}$, the roots of which will give multiple values to this symbol. But these values may be more readily found by means of the expansion of $f(x + h, y + ah)$; since by making $f(x + h, y + ah) = 0$, it is evident that h and ah will be corresponding increments of x and y in the equation $f(x, y) = 0$, and when these increments become infinitesimals, the symbol a will therefore represent the required values of $\frac{dy}{dx}$.

The expansion of $f(x + h, y + ah)$, given in art. (47), being equated with zero, omitting the first term $f(x, y)$, which $= 0$ by hypothesis, we obtain

$$0 = \frac{h}{1} \left\{ \left(\frac{du}{dx}\right) + \left(\frac{du}{dy}\right) a \right\}$$

$$+ \frac{h^2}{1.2} \left\{ \left(\frac{d^2 u}{dx^2}\right) + 2\left(\frac{d^2 u}{dx\,dy}\right) a + \left(\frac{d^2 u}{dy^2}\right) a^2 \right\}$$

$$+ \frac{h^3}{1.2.3} \left\{ \left(\frac{d^3 u}{dx^3}\right) + 3\left(\frac{d^3 u}{dx^2\,dy}\right) a + 3\left(\frac{d^3 u}{dx\,dy^2}\right) a^2 + \left(\frac{d^3 u}{dy^3}\right) a^3 \right\}$$

&c.   &c.   &c.

which may be made complete in any number of terms by replacing x and y by $x + \theta h$ and $y + \theta a h$ in the last term, where $\theta < 1$.

Now if particular values of x and y give $\left(\frac{du}{dx}\right) = 0$, $\left(\frac{du}{dy}\right) = 0$, the first term of this equation will disappear; and hence by stopping the series at the second term and dividing by the $\dfrac{h^2}{1.2}$, we get an equation determining the value of $a = \dfrac{\Delta y}{\Delta x}$ for all values of h, and finally, making $h = 0$, the $x + \theta h$, $y + a\theta h$ become simply x, y, and we obtain, for determining the continuous value of a, the equation

$$0 = \left(\frac{d^2 u}{dx^2}\right) + 2\left(\frac{d^2 u}{dx\, dy}\right) a + \left(\frac{d^2 u}{dy^2}\right) a^2,$$

a quadratic, which will therefore give two values for $a = \dfrac{dy}{dx}$.

If, however, for the same values of x and y, also

$$\left(\frac{d^2 u}{dx^2}\right) = 0, \qquad \left(\frac{d^2 u}{dx\, dy}\right) = 0, \qquad \left(\frac{d^2 u}{dy}\right) = 0,$$

then the first and second terms of the preceding equation will disappear, and hence stopping the series with the third term and, as before, dividing by the $\dfrac{h^3}{1.2.3}$ and afterwards making $h = 0$, we get

$$0 = \left(\frac{d^3 u}{dx^3}\right) + 3\left(\frac{d^3 u}{dx^2 dy}\right) a + 3\left(\frac{d^3 u}{dx\, dy^2}\right) a^2 + \left(\frac{d^3 u}{dy^3}\right) a^3,$$

a cubic equation, which will therefore determine three values for $a = \dfrac{dy}{dx}$.

Should the partial differential coefficients simultaneously vanish for still higher orders, the same process may be extended by including additional terms of the preceding form of development; but it will be unnecessary to do so here, as the general law of the successive terms is obvious, and these higher orders of multiple values do not often occur. It will be observed that the numerical coefficients of any order are those of the binomial theorem.

Example.— Given $y^3 - 7x^2 y - 6 x^3 + x^4 = 0$, to find the values of $\dfrac{dy}{dx}$, corresponding to $x = 0$ and $y = 0$.

When $x = 0$, $y = 0$, we have, by partial differentiation,

$$\left(\frac{du}{dx}\right) = -14\,xy - 18\,x^2 + 4\,x^3 = 0,$$

$$\left(\frac{du}{dy}\right) = 3\,y^2 - 7x^2 = 0\,;$$

$$\left(\frac{d^2u}{dx^2}\right) = -14\,y - 36\,x + 12\,x^2 = 0,$$

$$\left(\frac{d^2u}{dx\,dy}\right) = -14\,x = 0, \quad \left(\frac{d^2u}{dy^2}\right) = 6\,y = 0\,;$$

$$\left(\frac{d^3u}{dx^3}\right) = -36 + 24\,x = -36, \quad \left(\frac{d^3u}{dx^2\,dy}\right) = -14,$$

$$\left(\frac{d^3u}{dx\,dy^2}\right) = 0, \quad \left(\frac{d^3u}{dy^3}\right) = 6\,;$$

$$\therefore\ 0 = -36 - 42\,a + 6\,a^3, \quad \text{or } a^3 - 7\,a - 6 = 0,$$

the three roots of which are $a = 3$, -1 and -2; and these are therefore the required multiple values of $\frac{dy}{dx}$ when $x = 0$, $y = 0$.

(62.) The multiple values of a differential coefficient, which takes the form $\frac{0}{0}$, may be more simply and expeditiously determined algebraically in the following manner:

If the particular values of the variables be $x = a$, $y = b$, first transform the given function $f(x, y)$ by substituting $x' + a$, $y' + b$ respectively for x and y, so as to get the equivalent function in which the value of $\frac{dy'}{dx'}$ is to be obtained for $x' = 0$, $y' = 0$.

This last function being arranged in the ascending order of degree, with respect to the variables x', y', let it be denoted by

$$[x', y']_l + [x', y']_{l+m} + [x', y']_{l+m+n} + \&c. = 0,$$

where $[x', y']_l$ is supposed to comprise all the homogeneous terms of the least degree l with respect to x' and y', $(x', y')_{l+m}$ the homogeneous terms of the next higher degree $l + m$, &c. As these functions are homogeneous, it is evident that

$$\frac{[x', y']_l}{x'^l} = \left[1, \frac{y'}{x'}\right]_l ; \quad \frac{[x', y']_{l+m}}{x'^{l+m}} = \left[1, \frac{y'}{x'}\right]_{l+m}, \quad \&c.,$$

which will now represent algebraical functions of $\frac{y'}{x'}$. Hence, dividing the preceding equation by x', the result may be thus expressed :

$$\left[1, \frac{y'}{x'}\right]_l + x'^m \left[1, \frac{y'}{x'}\right]_{l+m} + x'^{m+n} \left[1, \frac{y'}{x'}\right]_{l+m+n} + \&c. = 0.$$

This equation, which must necessarily be true generally, determines $\frac{y'}{x'}$ as a function of x'. Now, when $x' = 0$, $y' = 0$, the continuous value of $\frac{y'}{x'}$ is obviously $\frac{dy'}{dx'}$ or $\frac{dy}{dx}$; and there-fore, making $x' = 0$ and replacing $\frac{y'}{x'}$ by $\frac{dy}{dx}$, the equation for determining this is

$$\left[1, \frac{dy}{dx}\right]_l = 0.$$

Hence the equation for determining the required values of $\frac{dy}{dx}$ is to be found by simply retaining only the homogeneous terms of least dimensions with respect to the variables, then dividing the same by a power of x' of equal dimensions, and finally replacing $\frac{y'}{x'}$ by $\frac{dy}{dx}$. The accuracy of the result will evidently not be affected, should the function, which comprises the terms of least dimensions, at the same time involve terms of higher dimensions that do not admit of convenient separa-tion, as these will finally vanish on making $x' = 0$, $y' = 0$.

This general rule will be found to apply with remarkable brevity and facility.

Example. — Take that given in the last article, viz. $y^3 - 7x^2y - 6x^3 + x^4 = 0$ to find the values of $\frac{dy}{dx}$ when $x = 0$, $y = 0$. Since the particular values of the variables are already $x = 0$, $y = 0$, the equation does not require any preliminary change. The first three terms are homogeneous and of the third degree, with respect to the variables; but the last term being of the fourth and therefore of a higher degree must be rejected. Hence, dividing $y^3 - 7x^2y - 6x^3$ by x^3 and replacing $\frac{y}{x}$ by $\frac{dy}{dx}$, we obtain

$$\left(\frac{dy}{dx}\right)^3 - 7\left(\frac{dy}{dx}\right) - 6 = 0,$$

the three roots of which are the values of $\left(\frac{dy}{dx}\right)$ as before found.

CHAPTER VI.

MAXIMA AND MINIMA.

(63.) The value of a function is a *maximum* if *less* values obtain when the variable is supposed to increase or decrease by small quantities.

The value is a *minimum* if *greater* values obtain when the variable is supposed to increase or decrease by small quantities.

A *maximum* value of a function is therefore *greater* and a *minimum* value is *less* than the values which immediately precede and follow it; and thus the relative analytical application of the terms *maxima* and *minima* has reference only to the values of the function which are immediately adjacent to the values so designated.

E

The same circumstances or conditions may recur for different values of the variable, and thus a function may admit of several maxima and minima, and the extreme values of these will obviously be the maximum and minimum values of the function in the absolute sense of the terms.

In some cases, however, the value of a function either always increases or always decreases when the variable is supposed to increase, and it therefore does not admit of an ordinary maximum or minimum according to the preceding definition.

1. *Functions of One Variable.*

(64.) Let $u = f(x)$ be a function of a variable x, and let it be required to find the particular values of the variable when the function is a maximum or a minimum.

Supposing the value of x to change by a small quantity h, if $f(x)$ be a maximum we must have $f(x) > f(x + h)$, and if $f(x)$ be a minimum we must have $f(x) < f(x + h)$, and these relations must be maintained whether h be positive or negative. Therefore, as h passes from $-$ to $+$, the value of the function $f(x)$ will be

$$\left.\begin{array}{l}\text{a maximum} \\ \text{a minimum} \\ \text{neither}\end{array}\right\} \text{ when } f(x + h) - f(x) \left\{\begin{array}{l}\text{continues to be negative,} \\ \text{continues to be positive,} \\ \text{changes its sign.}\end{array}\right.$$

But, art. (45),

$$f(x + h) - f(x) = h f'(x + \theta h).$$

If the first derived function $f'(x)$ have a finite value, it is evident that h may be taken so small that $f'(x + \theta h)$ shall not change its algebraic sign when that of h changes. As this value of $f(x + h) - f(x)$ will then have different signs, according to the sign of h, the function $f(x)$ will in such case be neither a maximum nor a minimum.

The preceding conditions of maxima and minima will require that h and $f'(x + \theta h)$ shall change sign simultaneously when h

passes through zero. But, art. (58), when a variable quantity changes its algebraic sign it must either pass through 0 or $\frac{1}{0}$. Therefore we must have $\frac{du}{dx} = f'(x) = 0$ or $\pm \infty$; and then supposing x, by increasing, to pass through its value, the function $f(x)$ will be

$$\left.\begin{array}{l} \text{a maximum} \\ \text{a minimum} \end{array}\right\} \text{ when } \frac{du}{dx} = f'(x) \text{ passes from } \left\{\begin{array}{l} + \text{ to} - \\ - \text{ to} +. \end{array}\right.$$

In the case $f'(x) = 0$, by extending Taylor's series to another term, we have

$$f(x + h) - f(x) = \frac{h^2}{1.2} f''(x + \theta h).$$

Here again, if $f''(x)$ be supposed not to vanish, the value of h may be taken so small that $f''(x + \theta h)$ shall not change sign when the sign of h is changed. As h^2 is necessarily positive the value of $f(x + h) - f(x)$ will have the same fixed algebraic sign as $f''(x + \theta h)$ or $f''(x)$; and therefore the function will be

$$\left.\begin{array}{l} \text{a maximum} \\ \text{a minimum} \end{array}\right\} \text{ when } \frac{d^2u}{dx^2} = f''(x) \text{ is } \left\{\begin{array}{l} \text{negative,} \\ \text{positive.} \end{array}\right.$$

Again, suppose that a value of x which makes $f'(x) = 0$ also causes several of the subsequent derived functions $f''(x)$, $f'''(x)$, &c. to vanish, and let $f^{(n)}(x)$ be the first that does not vanish. Then, art. (45),

$$f(x + h) - f(x) = \frac{h^n}{1.2 \ldots n} f^{(n)}(x + \theta h).$$

As $f^{(n)}(x)$ does not vanish, it is evident, as before, that a value may be assigned to h so small that $f^{(n)}(x + \theta h)$ shall not change its sign when that of h changes. The effect upon the sign of h^n will however depend upon whether the number n be odd or even. Thus we find,

If n be an *odd* number, $f(x)$ is neither a maximum nor a minimum, unless $f^{(n)}x$ passes through $\frac{1}{0}$.

If n be an *even* number,

$$f(x) \text{ is a } \left\{ \begin{array}{l} \text{maximum} \\ \text{minimum} \end{array} \right\} \text{ if } \frac{d^n u}{dx^n} = f^{(n)}x \text{ is } \left\{ \begin{array}{l} \text{negative,} \\ \text{positive.} \end{array} \right.$$

(65.) The nature of the preceding relations, which constitute the theory of maxima and minima of functions of one variable, may perhaps be made more familiar by the following simple considerations :

As the derived function $\frac{du}{dx} = f'(x)$ represents the limiting ratio of the increment of the function to that of the variable, and as a decrement is indicated by a negative increment, let the variable x be supposed to increase continuously ; then the value of the function $f(x)$ will increase when $f'(x)$ is positive and decrease when $f'(x)$ is negative.

But if $f(x)$ increases up to a certain value of x and afterwards decreases, it will evidently pass through a maximum value, and if it decreases and afterwards increases, it will pass through a minimum value. The function will therefore pass through a maximum or a minimum value whenever the value of the first derived function $\frac{du}{dx} = f'(x)$ passes from $+$ to $-$ or from $-$ to $+$ respectively.

After determining the values of x which make $f'(x) = 0$ and $\frac{1}{f'(x)} = 0$, this last simple criterion, which is that first obtained in art. (64), will generally be sufficient to distinguish the maxima and minima values, if any exist ; and then it will be unnecessary to proceed to any derived functions beyond $f'(x)$.

The process is also sometimes facilitated when the function admits of being reduced or simplified by first multiplying or dividing it by some constant, raising it to some power, taking the logarithm, or performing some other operation according to the particular form of the function under consideration, the only restriction being that this preparation of the function should not disturb the general relations as to corresponding maxima and minima.

(66.) The different cases specified in art. (64) may also be characterized geometrically by making the variable x the abscissa, and the function $f(x)$ the ordinate of a curve line, of which the equation is $y = f(x)$.

Fig. 1.

1. If for a value of x which makes $f'(x) = 0$, the value of $f''(x)$ is *negative*, or if the first of the successive derived functions that does not vanish be of an *even* order and its value *negative*, the corresponding value of the functional ordinate will be a *maximum* as represented in fig. 1.

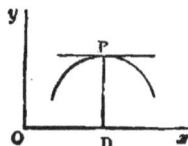

2. If for a value of x which makes $f'(x) = 0$, the value of $f''(x)$ is *positive*, or if the first of the successive derived functions that does not vanish be of an *even* order and its value *positive*, the corresponding value of the functional ordinate will be a *minimum* as represented in fig. 2.

Fig. 2.

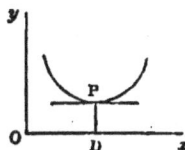

3. If for a value of x which makes $f'(x) = 0$, also $f''(x) = 0$, and the value of $f'''(x)$ is *positive*, or if the first of the successive derived functions that does not vanish be of an *odd* order and its value *positive*, or if the first of the derived functions that does not vanish be of an *even* order and its value passes through $\dfrac{1}{0}$ from $-\infty$ to $+\infty$, the corresponding value of the functional ordinate will be *neither* a maximum nor a minimum, and will be of the kind represented in fig. 3.

Fig. 3.

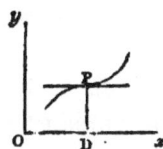

4. If for a value of x which makes $f'(x) = 0$, also $f''(x) = 0$, and the value of $f'''(x)$ is *negative*, or if the first of the successive derived functions that does not vanish be of an *odd* order and its value *negative*, or if

Fig. 4.

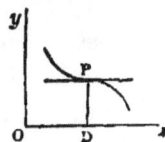

the first of the derived functions that does not vanish be of an *even* order and its value passes through $\dfrac{1}{0}$ from $+\infty$ to $-\infty$, the corresponding value of the functional ordinate will be *neither* a maximum nor a minimum, and will be of the kind represented in fig. 4.

5. If for a value of x which makes $\dfrac{1}{f'(x)} = 0$, the value of $f'(x)$, as x increases, passes from $+\infty$ to $-\infty$, or if for a value of x the first of the successive derived functions $f'(x)$, $f''(x)$, &c. that does not vanish is of an *odd* order and its value passes from $+\infty$ to $-\infty$, the corresponding value of the functional ordinate will be a *maximum* as represented in fig. 5 or fig. 1.

Fig. 5.

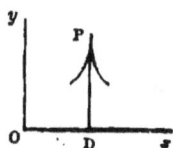

6. If for a value of x which makes $\dfrac{1}{f'(x)} = 0$, the value of $f'(x)$, as x increases, passes from $-\infty$ to $+\infty$, or if for a value of x the first of the derived functions $f'(x)$, $f''(x)$, &c. that does not vanish is of an *odd* order and its value passes from $-\infty$ to $+\infty$, the corresponding value of the functional ordinate will be a *minimum* as represented in fig. 6 or fig. 2.

Fig. 6.

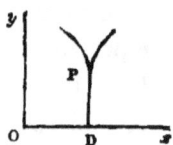

Example 1.—Divide a number a into two parts, such that their product shall be the greatest possible.

Let x be one of the parts, and $a - x$ the other; then $f(x) = x(a - x) = ax - x^2$ is required to be made a maximum; $\therefore f'(x) = a - 2x$ put $= 0$, gives $x = \frac{1}{2}a$. When x is less than $\frac{1}{2}a$ the value of $f'(x)$ is $+$, and when x exceeds $\frac{1}{2}a$ the value of $f'(x)$ is $-$; hence, when x passes through its value, $f'(x)$ passes through $+ \, 0 \, -$, which indicates that the value of the function first increases and then decreases, and therefore passes through a maximum, the number being then equally divided.

Example 2.—If $u = f(x) = 2x^3 - 9ax^2 + 12 a^2x - 4 a^3$; then

$$\frac{du}{dx} = f'(x) = 6x^2 - 18ax + 12 a^2 = 6 (x-a)(x-2a) = 0$$

gives $x = a$ and $x = 2a$. When x passes through the first of these values, $f'(x)$ passes through $+ 0 -$, which indicates a maximum, and when x passes through the second value, $f'(x)$ passes through $- 0 +$, which indicates a minimum. Therefore, when $x = a$, $f(x) = a^3$ a maximum, and when $x = 2a$, $f(x) = 0$ a minimum.

Ex. 3.—If $u = b + (x - a)^{\frac{4}{3}}$;

then $\frac{du}{dx} = f'(x) = \frac{4}{3} (x-a)^{\frac{1}{3}} = 0$ gives $x = a$, and as x passes through this value, $f'(x)$ passes through $- 0 +$, which indicates a minimum of the kind represented in fig. 2.

Ex. 4.—If $u = b + (x-a)^{\frac{5}{3}}$;

then $\frac{du}{dx} = f'(x) = \frac{5}{3} (x-a)^{\frac{2}{3}} = 0$ gives $x = a$. As x passes through this value, $f'(x)$ passes through $+ 0 +$ and does not change sign. The value of the function therefore first increases, then just ceases to increase, and again increases. It is hence neither a maximum nor a minimum, but of the character shown in fig. 3.

Ex. 5.—If $u = b + (x-a)^{\frac{2}{3}}$;

then $\frac{du}{dx} = f'(x) = \frac{2}{3} (x-a)^{-\frac{1}{3}}$, which $= \infty$ when $x = a$, and as x passes through this value, $f'(x)$ passes through $- \infty +$, which indicates a minimum of the kind represented in fig. 6.

Ex. 6.—Required the height (x) at which a light should be placed above a table so that a small portion of the surface of the table at a given horizontal distance (a) shall receive the greatest illumination from it.

If ϕ denote the angle under which the rays of light meet the given surface, the degree of illumination will vary as the sine of this angle directly and the square of the distance (r) inversely.

But $r^2 = a^2 + x^2$ and $\sin \phi = \dfrac{x}{r} = \dfrac{x}{\sqrt{a^2 + x^2}}$; $\therefore \dfrac{x}{(a^2 + x^2)^{\frac{3}{2}}}$

must be a maximum; or, taking the logarithm, the value of $\log x - \frac{3}{2} \log (a^2 + x^2)$ must be a maximum. Denoting this last function by u, we have

$$\frac{du}{dx} = \frac{1}{x} - \frac{3x}{a^2 + x^2} = \frac{a^2 - 2x^2}{x(a^2 + x^2)},$$

which $= 0$, when $x = a \sqrt{\frac{1}{2}}$, and as $\dfrac{du}{dx}$ passes through $+ \, 0 -$, the value of the function is then a maximum as required.

7. If $u = \dfrac{ax}{a^2 + x^2}$; then when $x = a$, $u = \frac{1}{2}$ a maximum, and when $x = -a$, $u = -\frac{1}{2}$ a minimum.

8. Of all rectangles of a given area, a square exhibits the least perimeter.

9. If $u = x^3 - 3ax^2 + 4a^3$; then $x = 0$ gives $u = 4a^3$ a maximum, and $x = 2a$ gives $u = 0$ a minimum.

10. If $u = \dfrac{\log x}{x}$; then when $x = e$, $u = \dfrac{1}{e}$ a maximum.

11. If $u = x^{\frac{1}{x^m}}$; then $x = e^{\frac{1}{m}}$ makes $u = e^{\frac{1}{me}}$ a maximum.

12. If $u = \dfrac{x}{(a + x)(b + x)}$;

then $x = \sqrt{ab}$ makes $u = \dfrac{1}{(\sqrt{a} + \sqrt{b})^2}$ a maximum.

13. If $u = \cos^3 x \sin x$; then $\cos^2 x = \frac{3}{4}$, $\sin^2 x = \frac{1}{4}$ give

$$u = \pm \frac{3}{16} \sqrt{3} \text{ a maximum and a minimum.}$$

II. *Functions of Two Variables.*

(67.) Let $u = f(x, y)$ be a function of two variables x and y. When the value of u is a maximum we must have $f(x, y)$

$> f(x + h, y + k)$, and when it is a minimum we must have $f(x, y) < f(x + h, y + k)$, and in either case this relation must remain unchanged whatever may be the algebraic signs of h and $k = ah$. Therefore, for all combinations of values and algebraic signs that can be given to the small quantities h and $k = ah$, if for brevity we put

$$f(x + h, y + ah) - f(x, y) = \delta u,$$

the value of the function u will be

$$
\left.\begin{array}{l}
\text{a maximum} \\
\text{a minimum} \\
\text{neither}
\end{array}\right\}
\text{ when } \delta u
\left\{\begin{array}{l}
\text{continues to be negative,} \\
\text{continues to be positive,} \\
\text{changes its sign.}
\end{array}\right.
$$

But, art. (47), we have

$$\delta u = h \left\{ \left(\frac{du}{dx}\right) + a \left(\frac{du}{dy}\right) \right\}_{\substack{x + \theta h \\ x + \theta ah.}}$$

When the value of this expression continues to be of the same algebraic sign, the value of the factor contained between the brackets, which corresponds to $x + \theta h$, $y + \theta ah$, must change sign with h, and this change of sign must occur when $h = 0$, or when $x + \theta h$, $y + a\theta h$ become x, y. Therefore, as the value of a is arbitrary, we must then have

$$\left(\frac{du}{dx}\right) = 0, \qquad \left(\frac{du}{dy}\right) = 0,$$

unless one or both of these partial differential coefficients should pass through the value $\frac{1}{0}$ with corresponding algebraic signs. These two equations or conditions will determine the particular values of the variables.

To ascertain further regarding the algebraic sign of the value of δu when $\left(\frac{du}{dx}\right) = 0$ and $\left(\frac{du}{dy}\right) = 0$, let the expansion of $f(x + h, y + ah)$, art. (47), be extended to another term; then, as the term of the first order in h now vanishes, we obtain

If the second differential coefficients do not severally vanish and their relative magnitudes be such that the value of

$$\left(\frac{d^2u}{dx^2}\right) + 2\,a\left(\frac{d^2u}{dx\,dy}\right) + a^2\left(\frac{d^2u}{dy^2}\right)$$

shall not vanish but continue of the same sign for all values of a, it is evident that h may be taken so small that the value of δu will always have a corresponding sign, which will not change with that of h. For brevity let this expression be denoted by

$$(A) + 2(c)\,a + (B)\,a^2\,;$$

then when $a = 0$ its value will be A, and, when the arbitrary quantity a, which is unrestricted in value, is made indefinitely great, its algebraic sign will be determined by that of B. The differential coefficients represented by A and B must therefore have like signs, and for all other values of a the expression must retain the same sign. By putting the expression under the equivalent form,

$$A\left\{\left(1 + \frac{c}{A}\,a\right)^2 + \frac{AB - c^2}{A^2}\,a^2\right\}$$

it becomes evident that it will necessarily have the same sign with the coefficient A when the value of $AB - c^2$ is positive, or $AB > c^2$; that is,

$$\left(\frac{d^2u}{dx^2}\right)\left(\frac{d^2u}{dy^2}\right) > \left(\frac{d^2u}{dx\,dy}\right)^2.$$

This is Lagrange's Condition of maxima and minima, and when it is satisfied the value of the function u will be

$$\begin{array}{l}\text{a maximum} \\ \text{a minimum}\end{array}\left\{\text{if } (A) = \left(\frac{d^2u}{dx^2}\right) \text{ is } \left\{\begin{array}{l}\text{negative,} \\ \text{positive.}\end{array}\right.\right.$$

If (A) and (B) or $\left(\frac{d^2u}{dx^2}\right)$ and $\left(\frac{d^2u}{dy^2}\right)$ have different signs, or if Lagrange's Condition be otherwise unsatisfied, the function is neither a maximum nor a minimum. Also if the values of x and y which make $\left(\frac{du}{dx}\right) = 0$, $\left(\frac{du}{dy}\right) = 0$ should happen to cause the second differential coefficients $\left(\dfrac{d^2u}{dx^2}\right)$, $\left(\dfrac{d^2u}{dx\,dy}\right)$, $\left(\dfrac{d^2u}{dy^2}\right)$ to vanish, it may be shown, as in art. (64), that a maximum or minimum value of the function will require that the first set of differential coefficients that do not vanish be of an even order.

III. *Functions of Three Variables.*

(68.) Let $u = f(x, y, z)$ be a function of three variables x, y, and z.

When u is a maximum $f(x, y, z) > f(x + h, y + k, z + l)$, and when it is a minimum $f(x, y, z) < f(x + h, y + k, z + l)$, where the symbols h, $k = ah$ and $l = \beta h$ denote small changes in the values of the variables. As in the last article, the values of x, y, z which maintain either of these relations must be found amongst the systems determined by the equations

$$\left(\frac{du}{dx}\right) = 0, \qquad \left(\frac{du}{dy}\right) = 0, \qquad \left(\frac{du}{dz}\right) = 0,$$

excepting, as before, the occurrence of infinite values.

If the second differential coefficients do not vanish, h may be taken so small that the value of

$$\delta u = f(x + h, y + ah, z + \beta h) - f(x, y, z)$$

shall have the same sign as the expression

$$\left(\frac{d^2u}{dx^2}\right) + \left(\frac{d^2u}{dy^2}\right)a^2 + \left(\frac{d^2u}{dz^2}\right)\beta^2 + 2\left(\frac{d^2u}{dy\,dz}\right)a\beta + 2\left(\frac{d^2u}{dz\,dx}\right)\beta$$
$$+ 2\left(\frac{d^2u}{dx\,dy}\right)a,$$

and not change its sign when that of h changes. For a maximum or a minimum therefore it will be essential that the value of this expression be either always negative or always positive, whatever values be given to the arbitrary quantities a and β, which are wholly unrestricted. To facilitate the determination of the requisite conditions amongst the coefficients, let the expression be more briefly denoted by

$$\epsilon = (A) + (B)\,a^2 + (C)\,\beta^2 + 2(a)\,a\beta + 2(b)\beta + 2(c)\,a$$

and by putting it under the equivalent form

$$A \left\{ \left(1 + \frac{b}{A}\beta + \frac{c}{A}a \right)^2 + \frac{AB - c^2}{A^2}a^2 + 2\,\frac{A\,a - bc}{A^2}\,a\beta \right.$$
$$\left. + \frac{AC - b^2}{A^2}\beta^2 \right\}$$

it is obvious that it will always have the same sign with the coefficient A, provided that the value of $(AB - c^2)\,a^2 + 2(A\,a - bc)\,a\beta + (AC - b^2)\,\beta^2$ be always positive, and this will be the case when $AB - c^2$ and $(AB - c^2)(AC - b^2) - (A\,a - bc)^2$ are both positive, or $AB > c^2$ and $(AB - c^2)(AC - b^2) > (A\,a - bc)^2$. There are therefore two conditions of maxima and minima, viz.

$$\left(\frac{d^2u}{dx^2} \right)\left(\frac{d^2u}{dy^2} \right) > \left(\frac{d^2u}{dx\,dy} \right)^2$$

$$\left\{ \left(\frac{d^2u}{dx^2} \right)\left(\frac{d^2u}{dy^2} \right) - \left(\frac{d^2u}{dx\,dy} \right)^2 \right\} \left\{ \left(\frac{d^2u}{dx^2} \right)\left(\frac{d^2u}{dz^2} \right) - \left(\frac{d^2u}{dx\,dz} \right)^2 \right\}$$
$$> \left\{ \left(\frac{d^2u}{dx^2} \right)\left(\frac{d^2u}{dy\,dz} \right) - \left(\frac{d^2u}{dx\,dz} \right)\left(\frac{d^2u}{dx\,dy} \right) \right\}^2.$$

* When *both* of these conditions are fulfilled, the function u will, as before, be

a maximum $\left.\begin{array}{l} \\ \\ \end{array}\right\}$ if $(A) = \left(\dfrac{d^2u}{dx^2} \right)$ is $\left\{\begin{array}{l} \text{negative,} \\ \text{positive.} \end{array}\right.$
a minimum

(69.) The conditions may be otherwise obtained in a symmetrical form, and the extreme value of ϵ determined as a maximum or minimum value of a function of two variables a, β. Thus we have

$$\left(\frac{d\epsilon}{da} \right) = 2\,(Ba + a\beta + c) = 0 \dots (1)$$

$$\left(\frac{d\epsilon}{d\beta} \right) = 2\,(C\beta + aa + b) = 0 \dots (2)$$

$$\left(\frac{d^2\epsilon}{da^2} \right) = 2\,B, \quad \left(\frac{d^2\epsilon}{d\beta^2} \right) = 2\,C, \quad \left(\frac{d^2\epsilon}{da\,d\beta} \right) = 2\,a.$$

---

* The first of these conditions is as essential as the second, although it is commonly neglected by writers on this subject.

Hence (67) if $BC > a^2$ the value of ϵ will be

a maximum ⎱ if A, B, and C are ⎰ negative,
a minimum ⎰   ⎱ positive;

so that if this value have the same sign as A, B, and C, *all* the values of ϵ will have the same sign. From equations (1) and (2) the values of a and β which determine this value of ϵ are

$$a = \frac{ab - Cc}{BC - a^2}, \qquad \beta = \frac{ac - Bb}{BC - a^2}.$$

For simplification, previous to the substitution of these values, multiply equation (1) by a, equation (2) by β, and add the results, and $Ba^2 + C\beta^2 + 2aa\beta + b\beta + ca = 0$. These terms being therefore omitted in the expression for ϵ, it becomes $\epsilon = A + b\beta + ca$, in which, now substituting the particular values of a, β, we get

$$\epsilon = \frac{ABC}{BC - a^2}\left(1 - \frac{a^2}{BC} - \frac{b^2}{CA} - \frac{c^2}{AB} + \frac{2\,abc}{ABC} \right) \; \dots \; (3).$$

When this extreme value of ϵ is of the same sign as A, B, and C, we have therefore the symmetrical condition

$$1 - \frac{a^2}{BC} - \frac{b^2}{CA} - \frac{c^2}{AB} + \frac{2\,abc}{ABC} > 0 \; \dots \; (4).$$

Also, putting

$$\cos^2\phi = \frac{a^2}{BC}, \qquad \cos^2\phi' = \frac{b^2}{CA}, \qquad \cos^2\phi'' = \frac{c^2}{AB} \; \dots \; (5),$$

the value of ϵ becomes

$$\epsilon = \frac{(A)}{\sin^2\phi}(1 - \cos^2\phi - \cos^2\phi' - \cos^2\phi'' + 2\cos\phi\cos\phi'\cos\phi'').$$

But if ϕ, ϕ', ϕ'' denote the sides of a spherical triangle, and ω, ω', ω'' the perpendiculars upon them from the opposite angles, this last expression, by spherics, is equivalent to

$$\epsilon = (A)\sin^2\omega = \left(\frac{d^2u}{dx^2}\right)\sin^2\omega;$$

$$\therefore \; \delta u = \frac{h^2}{1.2}\epsilon = \frac{h^2}{1.2}\left(\frac{d^2u}{dx^2}\right)\sin^2\omega,$$

which, for a given small increment h and arbitrary small increments k and l, represents the least possible value of δu when considered apart from its algebraic sign.

Similarly, for a given small increment k and arbitrary small increments l and h the least possible value of δu, or the value that approaches nearest to zero, is $\delta u = \dfrac{k^2}{1.2}\left(\dfrac{d^2u}{dy^2}\right)\sin^2\omega'$; and for a given increment l and arbitrary increments h and k, it is $\delta u = \dfrac{l^2}{1.2}\left(\dfrac{d^2u}{dz^2}\right)\sin^2\omega''$.

We also here conclude that the conditions of maxima or minima, with respect to the value of the function u, will be definitely indicated by the values of the angles ϕ, ϕ', ϕ'' given by equations (5). These conditions will be :

1. That the values of the angles be real.

2. That their relative magnitudes be such as to admit of being made the sides of a spherical triangle, which will simply require the value of each of them to be less than half their sum.

For functions of two variables there will be only one angle ϕ, and the analogous condition will only require that the value of this angle be real. Also the values of δu nearest to zero for a given value of h with k arbitrary and for a given value of k with h arbitrary will then be $\delta u = \dfrac{h^2}{1.2}\left(\dfrac{d^2u}{dx^2}\right)\sin^2\phi$ and $\delta u = \dfrac{k^2}{1.2}\left(\dfrac{d^2u}{dy^2}\right)\sin^2\phi.$

The form of the condition (4), for three variables, is equivalent to that first obtained, since $(AB - c^2)(AC - b^2) - (Aa - bc)^2 = A(ABC - Aa^2 - Bb^2 - Cc^2 + 2abc) > 0$, which divided by the positive factor A^2BC gives (4). Also when the values fulfil the condition (4) and any one of the three conditions $AB > c^2$, $BC > a^2$, $AC > b^2$, the other two will necessarily follow.

In conclusion, it may be as well to observe that the conditions and criteria of maxima and minima here investigated, though occasionally indispensable, are not often required, as the general circumstances are in most cases sufficiently indicated in the nature of the problem, and it is then only requisite to solve the equations $\left(\dfrac{du}{dx}\right) = 0,\ \left(\dfrac{du}{dy}\right) = 0,\ \left(\dfrac{du}{dz}\right) = 0$, for the determination of the variables.

CHAPTER VII.

PROPERTIES OF PLANE CURVES.

⁄1. *Quadrature and Rectification.*

(70.) The theory of plane curve lines forms a leading subject in Analytical Geometry of Two Dimensions, and the investigation of the various properties is generally found to be convenient and symmetrical when the positions are referred to rectangular coordinate axes.

In the annexed diagram let Ox, Oy represent the positive directions of the axes; then, $OD = x$, $DP = y$ being the two coordinates of the point P, the curve which is the locus of P is determined by an equation

$$y = \phi(x), \quad \text{or } f(x, y) = 0.$$

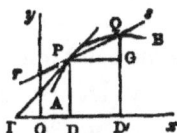

Suppose x and y to receive the increments Δx and Δy, and let the new coordinates $OD' = x + \Delta x$, $D'Q = y + \Delta y$ determine a second point Q, so that $DD' = PG = \Delta x$ and $GQ = \Delta y$. Then if A denote the function which expresses the value of the area contained between the ordinate, the curve, and the axis of x, the curvilinear area between the two ordinates DP, D'Q will geometrically represent the value of ΔA, and it is evident from the diagram that this value of ΔA will be comprised between the two rectangles $y \Delta x$ and $(y + \Delta y) \Delta x$, being greater than one and less than the other;

$\therefore\ \dfrac{\Delta A}{\Delta x}$ is comprised between y and $y + \Delta y$. Hence, proceeding to the continuous values at the limit when $\Delta x = 0$, we obtain

$$\frac{dA}{dx} = y, \quad \text{or } dA = y\,dx.$$

As this relation must correspond with the differentiation of A as a function of x, it is evident that the determination of A from it will be the inverse process to that of differentiation. This inverse process is called Integration, and is usually indicated by prefixing the symbol $\int$, thus

$$A = \int y\,dx.$$

The method of obtaining the value of this integral is the province of the Integral Calculus; and, when taken between given limits, it will express the area contained between the corresponding ordinates.

(71.) Again, let it be required to express, by means of infinitesimals, the area contained between the curve, two given ordinates y_0, y_m, and the axis of x.

Suppose a number $m - 1$ of equidistant ordinates y_1, y_2, $y_3 \ldots y_{m-1}$ to be inserted between them, and let dx be the common difference of the abscisses x_0, x_1, $x_2 \ldots x_m$. For brevity let $(y_0\,y_1)$ denote the portion of area contained between y_0, y_1, the axis of x and the curve, and the same for the other ordinates. Then it is evident that

$(y_0\,y_1)$ will be comprised between $y_0\,dx$ and $y_1\,dx$

$(y_1\,y_2)$,,   ,,   ,, $y_1\,dx$,, $y_2\,dx$

$(y_2\,y_3)$,,   ,,   ,, $y_2\,dx$,, $y_3\,dx$

    &c.   &c.   &c.

$(y_{m-1}\,y_m)$,,   ,,   ,, $y_{m-1}\,dx$,, $y_m\,dx$.

Hence, if

$$\Sigma y\,dx = y_0\,dx + y_1\,dx + y_2\,dx \ldots + y_{m-1}\,dx,$$

the sum of these relations proves that the total area $(y_0\,y_m)$ will be comprised between $\Sigma y\,dx$ and $\Sigma y\,dx + (y_m - y_0)dx$.

If we now suppose the number $m-1$ of intermediate ordinates to be increased without limit, dx and $(y_m - y_0)\, dx$ will decrease without limit, and therefore $\Sigma y\, dx$ will approximate to the proposed curvilinear area as its utmost limit; that is,

$$A = \Sigma y\, dx.$$

But we have seen that this curvilinear area is expressed by the integral $\int y\, dx$. Therefore

$$\int y\, dx = \Sigma y\, dx.$$

Hence it appears that every integral $\int y\, dx$ expresses that value to which $\Sigma y\, dx$ approximates as its ultimate limit, on increasing indefinitely the number of subdivisions dx, both being estimated between the same limiting values of x. This character of an integral presents to the mind a clear view as to the result of a process of integration, and the area of a curve offers the most simple geometrical representation of the process. When dx is taken indefinitely small so as to be considered as an infinitesimal, called an element of x, each of the terms $y\, dx$ of $\Sigma y\, dx$ is a similar element of the area; and we have shown that the nearer the values of these elements are taken to zero, the more accurately will they represent the relative changes of their respective primitive quantities, and the more accurately will a succession of them compose those quantities so as to form a continuous result. The idea of elements greatly facilitates our reasonings in the higher applications of the Differential and Integral Calculus, and gives to the mind the most ample scope in geometrical and physical researches, whilst a strict adherence either to the principle of derived functions or to what is usually called the theory of limits, which some authors rigidly contend for, would render many investigations exceedingly cramped, and others almost impossible.

(72.) If a right line rs which passes through the two points P and Q be supposed to revolve about the point P so that the intersection Q with the curve may proceed towards P, it has been shown, art. (9), that when the point Q arrives at the point P or when the distance PQ becomes an infinitesimal, the corresponding continuous position of the line *rs* will ultimately coincide with the tangent TP which touches the curve at the ´point P, and that the infinitesimal line PQ becomes then an element of the arc of the curve. These considerations are equivalent to that of conceiving the tangent to be a line which passes through two points of the curve that are infinitely near to each other. Let *s* denote the length of the arc from a given point in the curve to the point P; then will dx, dy, and ds symbolize the relative infinitesimal values of PG, GQ, and PQ. But $PQ^2 = PG^2 + GQ^2$;

$$\therefore \ ds^2 = dx^2 + dy^2$$

$$\text{and } s = \int \sqrt{dx^2 + dy^2} = \int dx \sqrt{1 + \frac{dy^2}{dx^2}}.$$

When y is known as a function of x, explicit or implicit, this expression serves to determine the length or rectification of the curve; but the inverse operation of integration, indicated by $\int$, will require the aid of the integral calculus.

II. *Tangent and Normal.*

(73.) Let ω denote the angle PTD or the inclination of the tangent with the axis of x; then, from what precedes, we have, as before deduced in art. (9),

$$\tan \omega = \frac{dy}{dx}.$$

If a, β be the coordinates of any point in the tangent PT, this gives

$$\frac{\beta - y}{a - x} = \frac{dy}{dx};$$

therefore the equation to the *tangent* is

$$\beta - y = \frac{dy}{dx}(a - x).$$

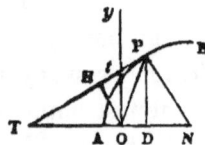

The *normal* PN being perpendicular to the tangent, if a', β'

be the coordinates of any of its points, its equation is hence

$$\beta' - y = -\frac{dx}{dy}(a' - x).$$

Hence if p denote the perpendicular O H from the origin upon the tangent and $p' = $ P H that upon the normal, we shall have

$$p = \frac{x\,dy - y\,dx}{ds}, \qquad p' = \frac{x\,dx + y\,dy}{ds}.$$

Also, if a'', β'' be the coordinates of any point in the line O H drawn through the origin perpendicular to the tangent, the equation to this line is

$$\beta'' = -\frac{dx}{dy}a''.$$

Again, since $\tan \omega = \dfrac{dy}{dx}$, and $ds^2 = dx^2 + dy^2$, we have

$$\cos \omega = \frac{dx}{ds}, \text{ and } \sin \omega = \frac{dy}{ds};$$

$$\therefore \text{ PT} = \text{tangent} = \frac{y}{\sin \omega} = \frac{y\,ds}{dy},$$

$$\text{PN} = \text{normal} = \frac{y}{\cos \omega} = \frac{y\,ds}{dx},$$

$$\text{DT} = \text{subtangent} = \frac{y}{\tan \omega} = \frac{y\,dx}{dy},$$

$$\text{DN} = \text{subnormal} = y \tan \omega = \frac{y\,dy}{dx}.$$

(74.) When the equation of the curve is of the form $u = f(x, y) = 0$, the differential elements dx, dy will be connected by the corresponding differential equation

$$\left(\frac{du}{dx}\right) dx + \left(\frac{du}{dy}\right) dy = 0.$$

Therefore the elements dx, dy, and ds will have the same mutual proportions as the respective quantities

$$\left(\frac{du}{dy}\right), \ -\left(\frac{du}{dx}\right) \text{ and } \sqrt{\left(\frac{du}{dx}\right)^2 + \left(\frac{du}{dy}\right)^2};$$

and by replacing them by these quantities the preceding relations, and any formulæ involving the ratios of the elements, will then become adapted to the case in which y is an implicit function of x.

The equation to the *tangent*, under this form, is thus

$$\left(\frac{du}{dx}\right)(a - x) + \left(\frac{du}{dy}\right)(\beta - y) = 0,$$

and it is therefore to be practically obtained by this simple rule : Differentiate the given equation of the curve, $u = f(x, y) = 0$, and write $a - x$, $\beta - y$ in place of dx and dy.

Also the equation of the normal is

$$\left(\frac{du}{dy}\right)(a' - x) - \left(\frac{du}{dx}\right)(\beta' - y) = 0.$$

Example.—The equation to an ellipse when referred to its centre and principal semidiameters a, b, is $\dfrac{x^2}{a^2} + \dfrac{y^2}{b^2} = 1$.

By differentiating, this gives $\dfrac{x}{a^2} dx + \dfrac{y}{b^2} dy = 0$;

$$\therefore \ \frac{dy}{dx} = -\frac{b^2 x}{a^2 y}, \qquad \frac{ds}{dx} = \frac{\sqrt{a^4 y^2 + b^4 x^2}}{a^2 y},$$

$$\frac{ds}{dy} = -\frac{\sqrt{a^4 y^2 + b^4 x^2}}{b^2 x},$$

$$\text{tangent} = \frac{y\sqrt{a^4 y^2 + b^4 x^2}}{b^2 x}, \quad \text{normal} = \frac{\sqrt{a^4 y^2 + b^4 x^2}}{a^2},$$

$$\text{subtangent} = -\frac{a^2 y^2}{b^2 x}, \text{ and subnormal} = -\frac{b^2}{a^2} x.$$

Also, the equation to the tangent is

$$\frac{x}{a^2}(a - x) + \frac{y}{b^2}(\beta - y) = 0, \quad \text{or } \frac{x}{a^2}a + \frac{y}{b^2}\beta = 1;$$

and the equation to the normal is

$$\frac{y}{b^2}(a' - x) - \frac{x}{a^2}(\beta' - y) = 0, \quad \text{or } \frac{a^2}{x}a' - \frac{b^2}{y}\beta' = a^2 - b^2.$$

III. *Asymptotes.*

(75.) Two curves or a curve and straight line are mutually asymptotic when they continually approach indefinitely nearer and nearer to each other, but do not meet at any finite distance. By an asymptote to a curve we generally understand a straight line, such that if it and the curve be indefinitely continued they will thus continually approach each other but never meet. It may therefore be considered as a determinate tangent to the curve when the point of contact is removed to an infinite distance.

The position of the tangent to the curve is geometrically determined when the intercepts OT, Ot of the coordinate axes are known.

In the equation of the tangent, art. (73), make $\beta = 0$, and we shall find the intercept of the axis of x, between the origin and the tangent, to be *

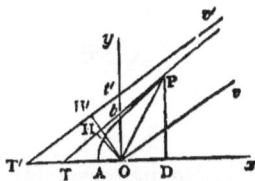

$$a_0 = \text{OT} = x - \frac{y\,dx}{dy} = \frac{x\,dy - y\,dx}{dy}.$$

Also, by making $a = 0$ we similarly find the corresponding intercept of the axis of y to be

$$\beta_0 = \text{O}t = y - \frac{x\,dy}{dx} = -\frac{x\,dy - y\,dx}{dx}.$$

---

* In the diagram, OT being in the contrary direction to Ox must be accounted a *negative* quantity, and equal to O D − D T.

If, when $x = \infty$ or $y = \infty$, either of these values of a_0 and β_0 should be finite, the curve will have one or more asymptotes which will thence be determined.

When a_0 is *infinite* and β_0 *finite* the asymptote is parallel to the axis of x.

When a_0 is *finite* and β_0 *infinite* the asymptote is parallel to the axis of y.

When a_0 and β_0 are *both finite* the asymptote passes through the two determined points T, t.

When the values of a_0 and β_0 are *both* $= 0$ the asymptote passes through the origin, and its direction will be determined by the value of $\frac{y}{x}$ when $x = \infty$ or $y = \infty$.

But when the values of a_0 and β_0 are *both of them infinite*, the tangent is at an infinite distance from the origin, cannot be constructed, and is not an asymptote.

The asymptotic branches of the curve will, with few exceptions, be analogous to one or other of the forms exhibited in the annexed diagrams, and will only differ with respect to relative situation.

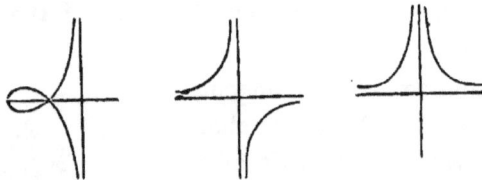

These diagrams, for example, may be considered to represent the general features of the respective curves determined by the equations

$$y = x \sqrt{\frac{a + x}{b - x}}, \quad y = -\frac{a^2}{x}, \text{ and } y = \frac{a^3}{x^2}.$$

When the axes of coordinates or lines parallel to them are asymptotes to a curve, the circumstance will at once be indicated as follows :

If, when $y = 0$, $x = \infty$, the axis of x is an asymptote; **and** if, when $x = 0$, $y = \infty$, the axis of y is an asymptote. Such is the case with the curve whose equation is $xy = a^2$.

If, when $y = b$, $x = \infty$, a line parallel to the axis of x, at the distance $y = b$, is an asymptote; and if when $x = a$, $y = \infty$, a line parallel to the axis of y, at the distance $x = a$, is an asymptote. Such is the case when the equation is $xy - ay - bx = 0$.

In other cases the position of the asymptotic tangent, if any such exist, will be ascertained by determining as before the values of the intercepts a_0 and β_0.

(76.) The practical calculation of the values of a_0, β_0 and of the equation to the asymptote may be considerably facilitated by putting the expressions under the following form:

$$a_0 = \frac{d\left(\dfrac{x}{y}\right)}{d\left(\dfrac{1}{y}\right)}, \qquad \beta_0 = \frac{d\left(\dfrac{y}{x}\right)}{d\left(\dfrac{1}{x}\right)}.$$

Now since $\dfrac{y - \beta}{x - a} = \dfrac{dy}{dx}$, where a, β are the coordinates of any point whatever in the tangent, if when $x = \infty$, $y = \infty$ this tangent be an asymptote and pass at a finite distance from the origin, this point can be taken so that a and β shall be both finite, and the relation then gives $\dfrac{y}{x} = \dfrac{dy}{dx}$. Let therefore $\dfrac{y}{x} = t$ and $\dfrac{1}{x} = v$; then $\beta_0 = \dfrac{dt}{dv}$, and the equation to the tangent when it becomes an asymptote is $y = \beta_0 + \dfrac{dy}{dx}x = \beta_0 + tx$. Hence the following easy rule:

In the given equation of the curve substitute $x = \dfrac{1}{v}$ and $y = \dfrac{t}{v}$, and, after reducing the equation so obtained in t and v, determine from this equation the values of t, and $\beta_0 = \dfrac{dt}{dv}$

when v is made to vanish; then, if the value of β_0 be finite, the equation to the required asymptote is

$$y = t_0 x + \beta_0.$$

If by making $t = \infty$ we obtain a finite corresponding value of v, this will determine an asymptote parallel to the axis of y at the distance $x = \dfrac{1}{v}$.

Example 1.—Let the equation to the curve be $xy - ay - bx = 0$; then substituting $\dfrac{1}{v}$ and $\dfrac{t}{v}$ for x and y, and reducing, we obtain

$$t - avt - bv = 0, \quad \beta = \frac{dt}{dv} = \frac{at + b}{1 - av}.$$

Therefore, making $v = 0$, we get $t_0 = 0$ and $\beta_0 = b$, and the equation of the asymptote is $y = b$, indicating that it is parallel to the axis of x at this distance.

By making $t = \infty$ we get $v = \dfrac{1}{a}$; $\therefore x = a$ is another asymptote and is parallel to the axis of y.

Example 2.—Let $y^3 + x^3 - axy = 0$; then substituting as before we get

$$t^3 + 1 - atv = 0, \quad \beta = \frac{dt}{dv} = \frac{at}{3t^2 - av}.$$

Hence making $v = 0$ we obtain $t_0 = -1$ and $\beta_0 = -\dfrac{a}{3}$, and the equation to the required asymptote is therefore

$$y = -x - \frac{a}{3}.$$

3. The curve $(x + 1)\, y = (x - 1)\, x$ has an asymptote determined by the equation $y = x - 2$.

4. The curve $y^3 - ax^2 + x^3 = 0$ has an asymptote determined by $y = \dfrac{a}{3} - x$.

5. The curve $y^3 - 2xy^2 + x^2 y = a^3$ has two asymptotes,

viz. the axis of x and the line $y = x$, which makes equal angles with the coordinate axes.

6. The curve $xy^2 - y = x^3 + 2ax^2 + bx + c$ has three asymptotes, viz. the axis of y and the two lines $y = x + a$ and $y = -x - a$.

IV. *Circle of Curvature.*

(77.) A tangent to a curve may be conceived to be a line drawn through two of its points which are indefinitely near to each other; and these points being considered as the extremities of a differential element of the curve, it is evident that the first differentials of the coordinates which appertain to the tangent will correspond with those of the curve at the point of contact.

Similarly, the *circle of curvature* or the *osculating circle* may be conceived to be that circle which passes through three consecutive points of the curve which are indefinitely near to each other, the position and magnitude of a circle being determined when three of its points are known.

These three points being considered as the extremities of two successive differential elements of the curve, it is evident that both the first and second differentials of the coordinates which belong to the circle and curve must correspond at the point of contact.

Let x'', y'' be the coordinates of the centre of the circle, and $x - x''$, $y - y''$ will be the two lines drawn from it respectively parallel to x and y and terminating in the circumference at the point of contact; hence, denoting its radius by ρ, its equation is

$$(x - x'')^2 + (y - y'')^2 = \rho^2.$$

Now since this circle corresponds with the curve at two other points contiguous to the point of contact, we may differentiate twice and consider the first and second differential of the ordinates x, y as agreeing with those of the curve.

Hence differentiating, observing that in proceeding to these points x'', y'' remain invariable, we get

$$dx\,(x-x'') + dy\,(y-y'') = 0,$$

$$d^2x\,(x-x'') + d^2y\,(y-y'') + ds^2 = 0\,;$$

where $ds^2 = dx^2 + dy^2$, art. (72), s denoting the length of the curve. The first of these two equations requires the centre of the circle to be situated in the normal, and the second completes the determination of its position. Thus, from the two equations we deduce

$$x-x'' = \frac{-\,dy\,ds^2}{dy\,d^2x - dx\,d^2y}, \quad y-y'' = \frac{dx\,ds^2}{dy\,d^2x - dx\,d^2y}.$$

Therefore, substituting these values in the equation $\rho^2 = (x-x'')^2 + (y-y'')^2$, we find

$$\rho = \frac{ds^3}{dy\,d^2x - dx\,d^2y}.$$

Having proceeded on the principle of general differentiation in obtaining this expression for the radius of curvature, we may hereafter assume an independent variable at pleasure. If we consider the axis of x to be horizontal, the value of the radius will be *positive* when the convex side of the curve is presented *upwards,* and it will be *negative* when the convex side of the curve is presented *downwards.*

(78.) The value of the radius of curvature may be otherwise determined by conceiving the centre of the circle to be the intersection of two normals drawn from two points which are indefinitely near to each other. Let P R, P'R be two consecutive normals meeting in R, the centre of curvature, the element PP' of the curve being ds. Let also two tangents be supposed to be drawn at P and P', the former making an angle ω with the axis of x. Then, as ω is decreasing, the angle included by the tangents will be $-d\omega$, and this must evidently be the same as that included by the normals.

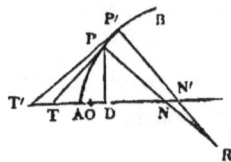

We have thus $PR = P'R = \rho$, $PP' = ds$, and the angle $PRP' = -d\omega$.

$$\therefore \ -\rho \, d\omega = ds$$

and
$$\rho = -\frac{ds}{d\omega}.$$

But, art. (73), $\tan \omega = \dfrac{dy}{dx}$; and hence, art. (29),

$$d\omega = \frac{d \tan \omega}{1 + \tan^2 \omega} = \frac{d\,\dfrac{dy}{dx}}{1 + \dfrac{dy^2}{dx^2}} = \frac{dx^2 . d\,\dfrac{dy}{dx}}{ds^2}.$$

Therefore, by substitution,

$$\rho = -\frac{ds^3}{dx^2 . d\,\dfrac{dy}{dx}} = \frac{ds^3}{dy\,d^2x - dx\,d^2y}.$$

By making x the independent variable, or supposing dx to be constant, this becomes

$$\rho = -\frac{ds^3}{dx\,d^2y} = -\frac{\left(1 + \dfrac{dy^2}{dx^2}\right)^{\frac{3}{2}}}{\dfrac{d^2y}{dx^2}},$$

which is the formula mostly employed in calculating the radius of curvature. The *measure of the curvature* of the curve at P will be the reciprocal of this radius, or $\dfrac{1}{\rho}$, being the same as that of the circle.

Differentiating the equation $dx^2 + dy^2 = ds^2$, we have

$$dx\,d^2x + dy\,d^2y = ds\,d^2s;$$
$$\therefore \ 0 = (dx\,d^2x + dy\,d^2y)^2 - (ds\,d^2s)^2.$$

Adding $(dy\,d^2x - dx\,d^2y)^2$ to this, the result is

$$(dy\,d^2x - dx\,d^2y)^2 = ds^2\{(d^2x)^2 + (d^2y)^2 - (d^2s)^2\}$$

$$\therefore \ \rho = \frac{ds^2}{\sqrt{\{(d^2x)^2 + (d^2y)^2 - (d^2s)^2\}}},$$

and, making s the independent variable, this becomes

$$\rho = \frac{ds^2}{\sqrt{\{(d^2x)^2 + (d^2y)^2\}}},$$

which is a symmetrical form of expression for the radius of curvature.

Example 1.—Find the radius of curvature at any point in an ellipse whose equation is $\dfrac{x^2}{a^2} + \dfrac{y^2}{b^2} = 1$.

Making x the independent variable, we have

$$\frac{dy}{dx} = -\frac{b^2x}{a^2y}, \text{ and } \frac{d^2y}{dx^2} = -\frac{b^4}{a^2y^3};$$

$$\therefore \rho = \frac{(a^4y^2 + b^4x^2)^{\frac{3}{2}}}{(ab)^4}.$$

Example 2.—In the cycloid, taking the vertex as the origin of coordinates,

$$y = \sqrt{2ax - x^2} + a \text{ vers}^{-1} \frac{x}{a};$$

$$\frac{dy}{dx} = \sqrt{\frac{2a - x}{x}}, \quad \frac{d^2y}{dx^2} = -\frac{a}{x\sqrt{2ax - x^2}};$$

$$\therefore \rho = 2\sqrt{2a(2a - x)}.$$

Example 3.—In the parabola $y^2 = 4mx$,

$$\rho = 2\sqrt{\frac{(m + x)^3}{m}}.$$

Example 4.—In the rectangular hyperbola, referred to its asymptotes, $2xy = a^2$, $\rho = -\dfrac{r^3}{a^2}$, r being the line drawn from the origin to the point in the curve.

Example 5.—In the conjugate hyperbolas $\dfrac{x^2}{a^2} - \dfrac{y^2}{b^2} = \pm 1$,

$$\rho = \frac{(a^4y^2 + b^4x^2)^{\frac{3}{2}}}{(ab)^4}.$$

Example 6. — In the catenary $y = \dfrac{c}{2}\left(e^{\frac{x}{c}} + e^{-\frac{x}{c}}\right)$, $\rho = -\dfrac{y^2}{c}$.

Example 7. — In the hypocycloid $x^{\frac{2}{3}} + y^{\frac{2}{3}} = a^{\frac{2}{3}}$,

$$\rho = -3(axy)^{\frac{1}{3}}.$$

v. *Evolute and Involute.*

(79.) If we suppose the point P to pass continuously through every point of the curve, the corresponding positions of the centre R of curvature will trace out another curve. This curve, which is the locus of the point R, is denominated the *evolute* of the proposed curve, and conversely the proposed curve is its involute. If the normal PR be supposed to move along with the point P, it is evident that the locus of the consecutive intersections R will be that curve to which the normal is always a tangent. This is rendered still further evident by considering it inversely: thus, by supposing a tangent to roll over a curve line, its successive indefinite intersections will obviously be the points of contact and therefore trace out the same curve. Hence a tangent drawn to the evolute at any point coincides with the radius of the osculating circle drawn to the point of contact. The equation of this tangent, art. (73), gives

$$dy''(x - x'') - dx''(y - y'') = 0.$$

Differentiate the equation

$$(x - x'')^2 + (y - y'')^2 = \rho^2,$$

supposing x'', y'', and ρ to vary, and we have

$$(dx - dx'')(x - x'') + (dy - dy'')(y - y'') = \rho\, d\rho\,;$$

but, x'', y'' appertaining to the normal of the curve at the point xy, we have by its equation

$$dx(x - x'') + dy(y - y'') = 0,$$

which rejected and the signs changed, we get

$$dx''(x - x'') + dy''(y - y'') = -\rho d\rho.$$

From this and the preceding equation to the tangent to the evolute we find

$$x - x'' = -\rho d\rho \frac{dx''}{ds''^2}, \qquad y - y'' = -\rho d\rho \frac{dy''}{ds''^2},$$

where $ds''^2 = dx''^2 + dy''^2$, s'' being the arc of the evolute from any given point.

These values of $x - x''$ and $y - y''$ being substituted in the equation $\rho^2 = (x - x'')^2 + (y - y'')^2$, we get

$$\rho^2 = \rho^2 \frac{d\rho^2}{ds''^2} \quad \text{or} \quad ds''^2 = d\rho^2;$$

$$\therefore \ ds'' = d\rho$$

$$\therefore \ s'' = \rho - \rho_0,$$

where ρ_0 is the radius of curvature corresponding to the given point from which s'' is estimated.

Hence the length of the arc of the evolute between any two points is equal to the difference between the radii of the corresponding osculating circles.

From this elegant property it follows that the original curve may be described by the unwinding of an inextensible thread from off the evolute. Thus if the normal or radius of curvature AQ be conceived to be a thread extending round the evolute QR, it is obvious that by unwinding this thread, keeping AQ always stretched, the point A will trace out the curve AB, and the unwound portion of the thread having passed from AQ to PR, the intercepted arc QR of the evolute will be equal to PR — AQ.

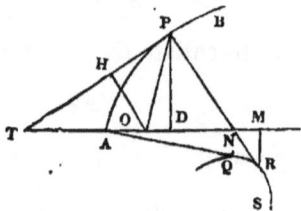

Considering the evolute as a primitive curve, its *involute* is thus described.

(80.) For the determination of the equation of the evolute

to any proposed curve we have, art. (77), the following expressions for the coordinates of the point R or of the centre of curvature, viz.

$$x'' = x + \frac{dy\, ds^2}{dy\, d^2x - dx\, d^2y} = x + \rho\frac{dy}{ds},$$

$$y'' = y - \frac{dx\, ds^2}{dy\, d^2x - dx\, d^2y} = y - \rho\frac{dx}{ds};$$

or, making x the independent variable,

$$x'' = x - \frac{dy\, ds^2}{dx\, d^2y} = x - \frac{dy}{dx}\cdot\frac{1 + \dfrac{dy^2}{dx^2}}{\dfrac{d^2y}{dx^2}}$$

$$y'' = y + \frac{ds^2}{d^2y} = y + \frac{1 + \dfrac{dy^2}{dx^2}}{\dfrac{d^2y}{dx^2}}.$$

By means of these and the equation of the curve AB, if the ordinates xy and their differentials admit of being eliminated an equation will thence be found expressing the relation between x'' and y'', and will be that of the evolute.

Let the equation of the evolute be given to find that of its involutes; then since $\rho = \rho_0 + s''$ and $d\rho = ds''$, the values of $x - x''$, $y - y''$, art. (79), give

$$x = x'' - (\rho_0 + s'')\frac{dx''}{ds''}, \qquad y = y'' - (\rho_0 + s'')\frac{dy}{ds''},$$

which being calculated in terms of x'' and y'', if these variables can be eliminated, the resulting equation in x and y will be the required equation to the involutes, ρ_0 being an arbitrary constant.

Example 1.—Determine the evolute of the Ellipse whose equation is

$$\frac{x^2}{a^2} + \frac{y^2}{b^2} = 1.$$

Taking x as the independent variable,

$$\frac{dy}{dx} = -\frac{b^2x}{a^2y}, \qquad \frac{d^2y}{dx^2} = -\frac{b^4}{a^2y^3};$$

$$x'' = \frac{a^2 - b^2}{a^4} x^3, \text{ and } y'' = -\frac{a^2 - b^3}{b^4} y^3;$$

$$\therefore \ x = a \left(\frac{a x''}{a^2 - b^2}\right)^{\frac{1}{3}}, \ y = -b \left(\frac{b y''}{a^2 - b^2}\right)^{\frac{1}{3}}, \text{ and by substitu-}$$

tion the required equation of the evolute is

$$(a x'')^{\frac{2}{3}} + (b y'')^{\frac{2}{3}} = (a^2 - b^2)^{\frac{2}{3}}.$$

Example 2.—The evolute to the parabola $y^2 = 4mx$ is the semicubical parabola $27 m y''^2 = 4 (x'' - 2m)^3$.

Example 3.—The evolute to the rectangular hyperbola $xy = a^2$ is $(x'' + y'')^{\frac{2}{3}} - (x'' - y'')^{\frac{2}{3}} = (4a)^{\frac{2}{3}}$.

Example 4.—The evolute to the hyperbola $\dfrac{x^2}{a^2} - \dfrac{y^2}{b^2} = 1$

is $(a x'')^{\frac{2}{3}} - (b y'')^{\frac{2}{3}} = (a^2 + b^2)^{\frac{2}{3}}.$

Example 5.—The evolute to the cycloid $y = \sqrt{2 a x - x^2}$ $+ a \text{ vers}^{-1} \dfrac{x}{a}$ is a cycloid equal to the original one, but in an inverse position.

vi. *Position of Convexity.*

(81.) As before, let ω denote the angle which the tangent to the curve at the point xy makes with the axis of x; then, art. (73),

$$\tan \omega = \frac{dy}{dx}.$$

For the purpose of conveniently expressing the relative positions, let the axis of x be considered to be horizontal, and that of y vertical, the positive direction of x being to the right hand and the positive direction of y being upwards. Then the tangent being supposed to be drawn in the positive direction with respect to the axis of x, its inclination (ω) with the horizontal will be

$$\left.\begin{matrix}\text{upwards}\\\text{downwards}\end{matrix}\right\} \text{ when } \tan \omega = \frac{dy}{dx} \text{ is } \begin{cases}\text{positive,}\\\text{negative.}\end{cases}$$

Now, when the curve at the point P, as in the diagram, has its convex side upwards, the angle ω thus estimated will evidently *decrease* as *x increases*; ∴ $\dfrac{d \tan \omega}{dx}$ will be *negative*.

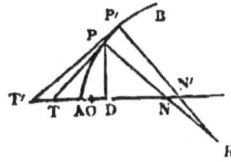

Also, when the convex side of the curve is downwards, the angle ω will *increase as x increases*, or $\dfrac{d \tan \omega}{dx}$ will be *positive*.

The position of *convexity* is therefore thus determined:

When $\dfrac{d^2y}{dx^2}$ is $\left\{ \begin{array}{l} \text{negative} \\ \text{positive} \end{array} \right\}$ it is presented $\left\{ \begin{array}{l} \text{upwards,} \\ \text{downwards.} \end{array} \right.$

In a similar manner the position of *convexity* with respect to the vertical will be determined by the algebraic sign of $\dfrac{d \tan \omega}{dy}$, or of $dy\ d \tan \omega$; and

When $\dfrac{dy}{dx} \cdot \dfrac{d^2y}{dx^2}$ is $\left\{ \begin{array}{l} \text{positive} \\ \text{negative} \end{array} \right\}$ it is $\left\{ \begin{array}{l} \text{to the right hand} \\ \text{to the left hand.} \end{array} \right.$

VII. *Points of Inflexion.*

(82.) When a curve is convex downwards, or in any other direction, and becomes afterwards convex in the opposite direction, it must have passed a point of contrary flexure in the vicinity of which the curve will resemble the middle turn of the letter s. In passing through one of these points, the second differential coefficient $\dfrac{d^2y}{dx^2}$, which determines the position of convexity upwards or downwards, must change its algebraic sign, and its value must therefore pass through 0 or $\dfrac{1}{0}$.

The condition for determining a point of contrary flexure or point of inflexion is therefore

$$\frac{d^2y}{dx^2} = 0 \text{ or } \infty.$$

I 5

If the value of $\dfrac{d^2y}{dx^2}$ at this point pass through $-0+$, the

inflexion will be of the character represented in diagram 1; and if it pass through $+0-$, it will be as exhibited in diagram 2. These two forms will represent all cases of inflexion if they are only placed in different positions with respect to the coordinate axes. It is also obvious that the value of the angle ω, which the tangent RS makes with the axis of x, will be a minimum in diagram 1, and a maximum in diagram 2.

Diagram 1.

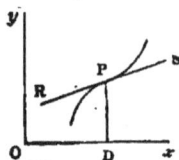

Diagram 2.

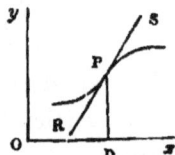

The expression, art. (78), for determining the radius of curvature ρ, contains $\dfrac{d^2y}{dx^2}$ in

the denominator. Therefore when $\dfrac{d^2y}{dx^2}$

passes through 0 and changes its sign, the value of the radius ρ will also change sign by passing through $\dfrac{1}{0}$. Hence the reason why the formula referred to expresses the value of ρ when the convex side of the curve is upwards, and gives to ρ a negative value when the convexity is downwards. Also as these radii are drawn in opposite directions, the centres of curvature being on opposite sides of the curve, this is in strict conformity with the usual geometrical interpretation of the symbols $+$ and $-$.

Example.—The Witch $xy = 2a(2ax - x^2)^{\frac{1}{2}}$ has two points of inflexion determined by $x = \dfrac{3a}{2}$, $y = \pm \dfrac{2}{3}a\sqrt{3}$.

(83.) *Note.*—When the equation to the curve is given in the implicit form $u = f(x, y) = 0$ the values of the differential coefficients, $\dfrac{dy}{dx}$, $\dfrac{d^2y}{dx^2}$, of y with respect to x, used in the preceding formulæ, arts. (75) to (82), will require some preliminary calculation. The consideration required for this may be obviated by expressing the formulæ in terms of the partial differential coefficients of the function $u = f(x, y)$. To effect this, the successive differentiation of the equation $u = 0$, art. (38), making x the independent variable and $d^2x = 0$, gives

$$\left(\frac{du}{dx}\right) dx + \left(\frac{du}{dy}\right) dy = 0,$$

$$\left(\frac{d^2u}{dx^2}\right) dx^2 + 2 \left(\frac{d^2u}{dx\,dy}\right) dx\,dy + \left(\frac{d^2u}{dy^2}\right) dy^2 + \left(\frac{du}{dy}\right) d^2y = 0;$$

or

$$\left(\frac{du}{dx}\right) + \left(\frac{du}{dy}\right) \frac{dy}{dx} = 0,$$

$$\left(\frac{d^2u}{dx^2}\right) + 2\left(\frac{d^2u}{dx\,dy}\right) \frac{dy}{dx} + \left(\frac{d^2u}{dy^2}\right) \frac{dy^2}{dx^2} + \left(\frac{du}{dy}\right) \frac{d^2y}{dx^2} = 0,$$

which are the relations connecting the values of $\frac{dy}{dx}$ and $\frac{d^2y}{dx^2}$ with those of the partial differential coefficients of u. Hence we obtain

$$\frac{dy}{dx} = -\frac{\left(\frac{du}{dx}\right)}{\left(\frac{du}{dy}\right)}.$$

$$\frac{d^2y}{dx^2} = -\frac{\left(\frac{d^2u}{dx^2}\right)\left(\frac{du}{dy}\right)^2 - 2\left(\frac{d^2u}{dx\,dy}\right)\left(\frac{du}{dx}\right)\left(\frac{du}{dy}\right) + \left(\frac{d^2u}{dy^2}\right)\left(\frac{du}{dx}\right)^2}{\left(\frac{du}{dy}\right)^3}.$$

The substitution of these values will accomplish the requisite transformation. For example, the expression for the radius of curvature, art. (78), becomes

$$\rho = \frac{\left\{\left(\frac{du}{dx}\right)^2 + \left(\frac{du}{dy}\right)^2\right\}^{\frac{3}{2}}}{\left(\frac{d^2u}{dx^2}\right)\left(\frac{du}{dy}\right)^2 - 2\left(\frac{d^2u}{dx\,dy}\right)\left(\frac{du}{dx}\right)\left(\frac{du}{dy}\right) + \left(\frac{d^2u}{dy^2}\right)\left(\frac{du}{dx}\right)^2},$$

which is necessarily symmetrical with respect to the co-ordinates.

The corresponding transformation of other formulæ is obvious and may be here left to the student.

VIII. *Multiple Points.*

(84.) A *multiple point* is a point in which two or more branches of a curve meet or intersect. If it is common to *two* branches of the curve it is called a *double point;* if it is the concourse of *three* branches it is called a *triple point,* &c.

At a multiple point there will be a tangent to each branch of the curve that passes through it, and therefore the differential coefficient $\frac{dy}{dx}$, which determines the position of the tangent, must admit of corresponding multiple values. In this case the expression for $\frac{dy}{dx}$, deduced from the equation of the curve, will take the indeterminate form $\frac{0}{0}$, and its multiple values may be obtained by either of the methods given in arts. (61) and (62).

Let $u = f(x, y) = 0$ be the equation to the curve; then, art. (61), the conditions for a multiple point will be

$$u = 0, \quad \left(\frac{du}{dx}\right) = 0, \quad \left(\frac{du}{dy}\right) = 0;$$

and if, for the values of x and y which simultaneously fulfil these equations, the second partial differential coefficients do not all vanish, the point will be double and the values of $a = \frac{dy}{dx}$ will be determined by the quadratic equation

$$\left(\frac{d^2u}{dx^2}\right) + 2a\left(\frac{d^2u}{dx\,dy}\right) + a^2\left(\frac{d^2u}{dy^2}\right) = 0.$$

For the convenience of abbreviation, let this be denoted by

$$(A) + 2(c)a + (B)a^2 = 0;$$

then the two values of a will be

$$a = \frac{c \pm \sqrt{c^2 - AB}}{B}.$$

We may hence, according to the nature of these roots of the quadratic, distinguish three classes of double points :

I. If the two roots or values of a be real and unequal, the two branches of the curve will take different directions, and the point will be a point of intersection or *real double point* as represented in diagrams 1 and 2. These and the following diagrams may be placed in any position with respect to the axes of coordinates.

Diagram 1.   Diagram 2.

II. If the values of a be equal, the two branches· of the curve will have a common tangent, and therefore also have mutual contact at the point under consideration.   In this case if the convexities of the two branches be situated on opposite sides, the contact will be external, as shown in diagram 3, and the point is called a point of *contact* of the *first kind* or point of *embrassement;* and if the convexities lie in the same direction the contact will be internal, as in diagram 4, and the point is then called a point of *contact* of the *second kind* or point of *osculation.*

Diagram 3.

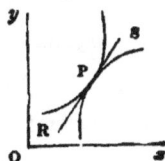

Diagram 4.

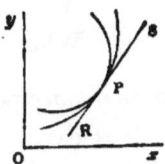

If, however, the value of $c^2 - AB$ under the radical, which vanishes at the point P, should change its sign and become negative on one side of the point, the corresponding value of a will be unreal, and therefore the two branches of the curve will be restricted to one side of the point, which is then denominated a *cusp.* As before, if the convexities of the two branches lie in contrary directions, the cusp is of the *first kind,* as shown in

diagram 5; and if the convexities are in the same direction
it is of the *second kind*, as shown in dia-

Diagram 5.

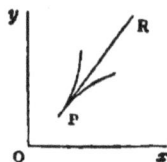

gram 6.

III. If the values of a be unreal, then no
real branch of the curve can pass through
or meet the proposed point, which, being
thus detached from its associated curve line,
is in such case called an *isolated* or *conjugate point.*

(85.) The analytical criteria for discrimi-
nating the character of a double point are
therefore as follows :

Diagram 6.

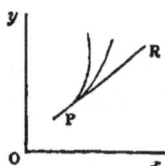

Let $u = 0,$ $\left(\dfrac{du}{dx}\right) = 0,$ $\left(\dfrac{du}{dy}\right) = 0$; then

I. When $\left(\dfrac{d^2u}{dx\,dy}\right)^2 - \left(\dfrac{d^2u}{dx^2}\right)\left(\dfrac{d^2u}{dy^2}\right) > 0,$

the point is an *intersection* of two branches of the curve and
is a *real double point.*

II. When $\left(\dfrac{d^2u}{dx\,dy}\right)^2 - \left(\dfrac{d^2u}{dx^2}\right)\left(\dfrac{d^2u}{dy^2}\right) = 0$; if > 0 for points
immediately preceding and following, it is a *contact* of two
branches; if of different signs at these points, it is a *cusp.*
The *contact* or *cusp* will be of the *first* or *second* kind
according as $\dfrac{d^2y}{dx^2}$ for the two branches has *different* signs or
the *same* sign.   If $\dfrac{d^2y}{dx^2} = 0$, this will indicate an *inflexion.*

III. When $\left(\dfrac{d^2u}{dx\,dy}\right)^2 - \left(\dfrac{d^2u}{dx^2}\right)\left(\dfrac{d^2u}{dy^2}\right) < 0,$ it is an *isolated*
or *conjugate* point.

It is easy to extend the process to higher orders of multiplicity.   If, for the values of x and y which fulfil the
equations $u = 0, \left(\dfrac{du}{dx}\right) = 0, \left(\dfrac{du}{dy}\right) = 0$;

also $\left(\dfrac{d^2u}{dx^2}\right) = 0, \left(\dfrac{d^2u}{dx\,dy}\right) = 0, \left(\dfrac{d^2u}{dy^2}\right) = 0$, and the third partial differential coefficients do not vanish, then the values of a will be the roots of the cubic equation

$$\left(\frac{d^3u}{dx^3}\right) + 3\,a\left(\frac{d^3u}{dx^2\,dy}\right) + 3\,a^2\left(\frac{d^3u}{dx\,dy^2}\right) + a^3\left(\frac{d^3u}{dy^3}\right) = 0.$$

If the three roots of this equation be real and unequal, the point will be an intersection of three branches or a *real triple point*, of which the point P in the annexed diagram, No. 7, is an example.

Diagram 7.

If two of the roots be equal, it will be a point of *contact* and *intersection;* if the three roots be equal, it will be a point of *double contact;* but if the equation contain a pair of unreal roots, then only one real branch of the curve passes through the point, and it is therefore in that case not a real triple point.

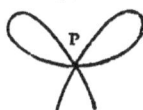

Should the point P be a *quadruple point*, as in diagram 8, the third partial differential coefficients will also vanish, and the values of a will be determined in like manner by an equation of the fourth degree.

Diagram 8.

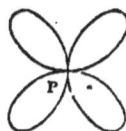

Since an algebraic equation of odd dimensions must necessarily have at least one real root, it is evident that a conjugate point can only occur when the degree of multiplicity is even.

(86.) An examination of the character of multiplicity of any proposed point of a curve may in general be more readily effected by a method analogous to that given in art. (62), for determining multiple values of $\frac{dy}{dx}$ when of the form $\frac{0}{0}$, and which we shall here repeat with a slight modification.

Let the coordinates of the point P be $x = a$, $y = b$; then if in the equation of the curve x and y be replaced by $a + x'$, $b + y'$, we shall have an equation in which x', y' are now the coordinates of any other point P' in the curve estimated from the proposed point P as a new origin. In this equation make

$y' = \beta x'$; then dividing throughout by the power of x' that may be common to the several terms, we shall obtain an equation

$$\phi(x', \beta) = 0,$$

in which β will denote $\dfrac{y'}{x'}$ or the tangent of the angle which the chord PP' makes with x', and when x' is made $= 0$ the corresponding values of β_0 given by this equation will evidently be those of $\dfrac{dy}{dx}$, and the number of such values will, as before, determine the multiplicity of the point.

Also, by giving to x' a small positive or a small negative value, we may ascertain the number and situation of the corresponding points P' in the immediate vicinity of P on either side.

Since $y' = \beta x'$ we have, by differentiating with x' as the independent variable,

$$\frac{dy'}{dx'} = \beta + x'\frac{d\beta}{dx'}, \qquad \frac{d^2y'}{dx'^2} = 2\frac{d\beta}{dx'} + x'\frac{d^2\beta}{dx'^2};$$

therefore at the point P, where $x' = 0$,

$$\frac{dy'}{dx'} = \beta_0, \qquad \frac{d^2y'}{dx'^2} = 2\left(\frac{d\beta}{dx'}\right)_0.$$

The first of these shows that the values of β when $x' = 0$ are those of $\dfrac{dy}{dx}$, as before stated; the second will determine the positions of convexity by art. (81) or the radii of curvature by art. (78) if required, the formula for the latter being

$$\rho_0 = -\frac{(1 + \beta_0^2)^{\frac{3}{2}}}{2\left(\dfrac{d\beta}{dx'}\right)_0}.$$

The nature of each separate branch of the curve may, however, be easily made known by comparing with β_0 the two values of β which correspond to small positive and negative values of x'. Thus, if $(\beta - \beta_0)\, x'$ continues to be positive, the convexity is evidently downwards; if it continue to be negative, the convexity is upwards; and if it change sign with x, the point is one of inflexion.

Example 1.—Let $x^4 - ax^2y + by^3 = 0$, and determine the nature of the point at the origin where $x = 0$, $y = 0$.

Here

$$\left(\frac{du}{dx}\right) = 4x^3 - 2axy = 0, \qquad \left(\frac{du}{dy}\right) = -ax^2 + 3by^2 = 0;$$

$$\left(\frac{d^2u}{dx^2}\right) = 12\,x^2 - 2\,ay = 0, \left(\frac{d^2u}{dx\,dy}\right) = -2ax = 0, \left(\frac{d^2u}{dy^2}\right) = 6by = 0;$$

$$\left(\frac{d^3u}{dx^3}\right) = 24x = 0, \left(\frac{d^3u}{dx^2\,dy}\right) = -2a, \left(\frac{d^3u}{dx\,dy^2}\right) = 0, \left(\frac{d^3u}{dy^3}\right) = 6b.$$

Therefore the equation for determining the values of $a = \dfrac{dy}{dx}$ is

$$-6aa + 6ba^3 = 0, \quad \text{or } ba^3 - aa = 0;$$

the roots of which are $a = 0$, and $a = \pm \sqrt{\dfrac{a}{b}}$, a   herefore the point is a real triple point similar to that shown in diagram 7.

Otherwise, the origin being already situated at the proposed point P, substitute $y = \beta x$, and $x^4 - a\,x^3\beta + b x^3 \beta^3 = 0$, which divided by x^3 gives $x - a\beta + b\beta^3 = 0$. Hence, at the origin, $-a\beta + b\beta^3 = 0$; $\therefore \beta = 0$ and $\beta = \pm \sqrt{\dfrac{a}{b}}$, and the point is a real triple point.

Example 2. — The equation being $ay^2 + b x^2 - x^3 = 0$, required the nature of the point at the origin.

Substitute βx for y and divide by x^2; then, $a\beta^2 + b - x = 0$; $\therefore \beta^2 = -\dfrac{b - x}{a}$, and at the origin, $x = 0$ and

$\beta_0 = \sqrt{-\dfrac{b}{a}}$, which being unreal, the point is detached from its curve, and is a conjugate point.

Example 3.—The curve $(ay-x^2)^2(a^2+x^2) - m^2a^2x^4 = 0$ passes through the origin; it is required to find the nature of this point.

Substitute, as before, $y = \beta x$; then, dividing by x^2, we get,

$$(a\beta - x)^2(a^2+x^2) - m^2a^2x^2 = 0; \quad \therefore \beta = \frac{x}{a} \pm \frac{mx}{\sqrt{a^2+x^2}}.$$

At the origin $\beta_0 = 0$, and, as the double values of β here merge into one, the two branches have mutual contact with the axis of x at this point. Differentiating the value of β we have also

$$\frac{d\beta}{dx} = \frac{1}{a} \pm \frac{ma^2}{(a^2+x^2)^{\frac{3}{2}}}, \quad \therefore \left(\frac{d\beta}{dx}\right)_0 = \frac{1 \pm m}{a}.$$

Therefore, if $m > 1$, the convexities lie in opposite directions and the contact is external; if $m < 1$, the contact is internal, or a point of osculation, and the two branches have their convexities presented downwards; and in either case the two radii of curvature are $\rho_0 = -\dfrac{a}{2(1 \pm m)}.$

Example 4.—The curve whose equation is $ax^2 + x^3 - by^2 = 0$ has a double point at the origin, and the directions of the branches are determined by $\beta_0 = \pm \sqrt{\dfrac{a}{b}}.$

Example 5.—The curve $(a^2-x^2)y^2 - (a^2+x^2)x^2 = 0$ has a double point at the origin, and $\beta_0 = \pm 1$, or the branches make equal angles with the axes of coordinates.

Example 6.—The Lemniscate $(x^2+y^2)^2 - a^2(x^2-y^2) = 0$ has a double point at the origin, and the branches make equal angles with the axes.

Example 7.—If $b(y-x)^2 - x^3 = 0$, the origin will be

a cusp of the first kind, the common tangent making equal angles with the axes.

Example 8.—If $x^5 + a^2 x^3 - b^3 y^2 = 0$, the origin will be a cusp of the first kind touching the axis of x.

Example 9.—In the Cissoid $y^2(2a - x) - x^3 = 0$, the origin is a cusp of the first kind also touching the axis of x.

Example 10.—If $(ay - ax - x^2)^2 - x^5 = 0$, the origin will be a cusp of the second kind, with the two convexities downwards, and the common tangent making equal angles with the coordinate axes; also the branches at this point will have the same centre of curvature, the common radius being $\rho_0 = -a\sqrt{2}$, so that the contact is of the second order.

Example 11.—The evolute to the ellipse, example 1, art. (80),

$$(ax)^{\frac{2}{3}} + (by)^{\frac{2}{3}} = (a^2 - b^2)^{\frac{2}{3}}$$

has four cusps of the first kind at the points

$$x = 0, \; y = \pm \frac{a^2 - b^2}{b}, \text{ and } y = 0, \; x = \pm \frac{a^2 - b^2}{a}.$$

IX. *Tracing of Curves.*

(87.) The equation of a curve being given, it is sometimes required to develop its particular structure, peculiarities of form, and general character. Such an investigation is usually called *discussing* or *tracing* a curve from its equation, and only requires the practical application of the preceding formulæ. It will be sufficient here to indicate the chief points that should engage attention.

I. If the equation be in the implicit form, it will be advisable, if practicable, to solve it with respect to one of the variables, provided the result be in a convenient form for calculation.

By first making $y = 0$ and then $x = 0$, we shall ascertain if the curve crosses the axes and the positions $(x_0, 0)$, $(0, y_0)$ of the points of intersection. Also, by assigning to one of the

variables a series of positive values from 0 to ∞, and of
negative values from 0 to −∞, and calculating the corresponding values of the other variable, we shall be enabled to follow
the course of the curve, and to discover if it has any infinite
branches.   In all these calculations both positive and negative
results should be carefully included, so as to obtain the complete branches of the curve.

II. Should the curve possess any infinite branches, ascertain
if they have asymptotes and determine their equations, and
thence their geometrical positions.

III. Determine the value of $\frac{dy}{dx}$, and from it deduce the maximum and minimum values of x and y, and the angles at which
the curve cuts the axes, &c.

IV. Determine the value of $\frac{d^2y}{dx^2}$ and thence the relative positions of convexity of the different branches, and the points of
inflexion if there be any.

V. Should the expression for $\frac{dy}{dx}$, for particular values of the
variables, become of the form $\frac{0}{0}$, determine the nature of the
corresponding multiple points.

Note.— In some cases the character of a curve can be
discussed with greater facility when its equation is transformed
into polar coordinates.   See the following Chapter.

X. *Envelopes.*

(88.) Let the equation to a system or family of curves be
denoted by

$$U = f(x, y, a) = 0,$$

where a is a variable parameter which is only constant for
each curve.   For each specific value of a the equation will be
that of a determinate curve ; and when a varies continuously

it will determine a continuous succession of curves, the position and character of each of which will differ but little from that which precedes it.

Let

$$U_0 = f(x, y, a) = 0,$$
$$U_1 = f(x, y, a + da) = 0,$$
$$U_2 = f(x, y, a + 2da) = 0,$$

be three consecutive curves in this series, and suppose P to be a point in which the curves U_0 and U_1 mutually intersect, and P' the corresponding point in which U_1 and U_2 intersect. Then, since the two points P, P' are both situated in the curve U_1, it is evident that the curve which is the locus of the points P will have the element of its arc, $PP' = ds$, coinciding with an equal element of the curve U_1. Therefore the curve traced by the intersection P will have contact with the entire family of curves U, and it is hence called the *envelope* of the system.

The envelope to the family of curves U is therefore to be found by determining the locus of the point of intersection of two consecutive curves taken indefinitely near to each other. Let x, y be the coordinates of the point of intersection P; then these coordinates will fulfil both of the equations $U = 0$, $U_1 = 0$. Hence, in passing from U to U_1, the point P will remain fixed and only a will vary, so that we must have

$$\left(\frac{dU}{da}\right) = 0.$$

We have thus the two equations

$$U = 0, \qquad \left(\frac{dU}{da}\right) = 0,$$

from which the variable parameter a being eliminated we shall obtain an equation involving x and y, the coordinates of the point P, which will be the equation to the envelope of the proposed curves U.

(89.) If the equation $U = f(x, y, a)$ be of the first degree in x and y, it will represent a system of straight lines; and if, as the parameter a varies continuously, the variable line be supposed to be in motion, the point P will obviously be the centre of instantaneous rotation; and its locus will be that curve to which the line is always a tangent. This may be made apparent by conceiving the envelope or the curve which is the locus of P to be represented by a rectilinear polygon of an indefinite number of sides, each of these sides at the same time representing an infinitesimal element ds of the curve. The sides produced will represent tangents to the curve, and the angular points will evidently be the intersections of consecutive tangents.

This property of a curve being generated by the ultimate intersections of a series of lines determined by a given law may be further instanced in the evolute to a curve. Since, art. (79), the normal drawn to a curve at any point is always a tangent to the evolute, it is evident that the evolute will be the envelope to all the normals, in the same way that a curve is the envelope to all its tangents.

Example 1.—Find the envelope to the system of lines determined by the equation $\dfrac{x}{a} + \dfrac{y}{\beta} = 1$, where a and β are variable parameters subject to the condition $a\beta = 4m^2$.

By differentiating the equations with respect to the parameters, we have

$$\frac{x}{a^2} da + \frac{y}{\beta^2} d\beta = 0,$$

$$\beta \, da + a \, d\beta = 0,$$

from which eliminating da, $d\beta$, we get $\dfrac{x}{a} = \dfrac{y}{\beta} = \dfrac{1}{2}$, or $a = 2x$, $\beta = 2y$. These substituted in $a\beta = 4m^2$, we have for the envelope the equation $xy = m^2$, which is that of a hyperbola referred to its asymptotes.

Example 2.—The equation to an ellipse being $\dfrac{x'^2}{a^2} + \dfrac{y'^2}{b^2} = 1$, that of the normal drawn through the point $x'y'$ is, example art. (74), $\dfrac{a^2 x}{x'} - \dfrac{b^2 y}{y'} = a^2 - b^2$; determine the envelope to all these normals.

The two variable parameters x', y' may be reduced to one by making $x' = a \cos a$, $y' = b \sin a$; then, putting $c^2 = a^2 - b^2$, we shall have

$$U = \frac{ax}{\cos a} - \frac{by}{\sin a} - c^2 = 0;$$

and, differentiating with respect to the variable parameter a,

$$\left(\frac{dU}{da}\right) = ax \frac{\sin a}{\cos^2 a} + by \frac{\cos a}{\sin^2 a} = 0.$$

From the latter equation, $\tan a = -\left(\dfrac{by}{ax}\right)^{\frac{1}{3}}$; and by substituting the corresponding values of $\cos a$, $\sin a$ in $U = 0$ and reducing we finally obtain

$$(ax)^{\frac{2}{3}} + (by)^{\frac{2}{3}} = (c^2)^{\frac{2}{3}},$$

which is the evolute to the ellipse, and agrees with the result before obtained in art. (80).

Example 3.—The envelope to the system of straight lines determined by the equation $y = ax + \dfrac{m}{a}$ is the parabola $y^2 = 4mx$.

Example 4.—The envelope to the system of circles $(x - m - a)^2 + y^2 = 4ma$ is also the parabola $y^2 = 4mx$.

Example 5.—If a straight line whose length is c slide with its extremities upon the axes of coordinates, its variable equation will be represented by $\dfrac{x}{c \cos a} + \dfrac{y}{c \sin a} = 1$; and the envelope, or curve to which the line is always a tangent, will be the hypotrochoid $x^{\frac{2}{3}} + y^{\frac{2}{3}} = c^{\frac{2}{3}}$.

Example 6.—The parabolas described by projectiles discharged, in vacuo, from a given point with a given velocity are included in the equation $4mv = 4max - (1 + a^2) x^2$; and the envelope to these is the parabola $x^2 = 4m (m - y)$.

CHAPTER VIII.

FORMULÆ FOR POLAR EQUATIONS, &C.

(90.) The system of representing positions by means of coordinates relative to fixed axes gives the greatest facility and the widest range to the applications of the analysis. It is on that account much employed in geometry, and almost exclusively in physics, to which in nearly every branch of inquiry it seems to be particularly adapted. In the geometry of curve lines, however, it is sometimes convenient to investigate the properties of certain curves from what is called the *polar equation*, and which is especially applicable to curves of the spiral kind.

A fixed indefinite right line Ox, originating at O, is called the *polar axis* or *prime radius;* the fixed point O is the pole or *origin ;* any right line OP drawn from the pole O to a variable point P is called the *radius vector* to that point, and its angle POx with the axis the *polar angle.*

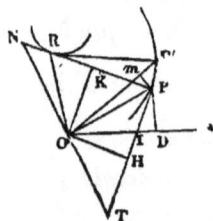

The radius vector OP is denoted by r, and the polar angle POx by θ; these evidently define the position of the point P, which may be symbolically designated the point $r\theta$.

The polar equation to a curve expresses a relation between r and θ, and is of the form

$$F(r, \theta) = c;$$

and, in most cases, r may be separated so as to give the explicit form

$$r = f(\theta),$$

F and f in most cases involving the polar angle θ under the form of trigonometrical functions.

The quantities r, θ being thus made subject to an equation, we shall have particular values of r for each successive value of θ; and hence the point P becomes restricted to a particular curve determined by the equation.

The perpendicular O H from the pole upon the. tangent being, as before, denoted by p, the equation to a curve is in some cases advantageously expressed in r and p.

(91.) *Polar Equivalents.*—By taking the axis of x for the polar axis, and the origin of the rectangular coordinates for the pole, we shall obviously have

$$x = r \cos \theta, \quad y = r \sin \theta;$$

and hence also, by differentiation,

$$dx = dr \cos \theta - r d\theta \sin \theta,$$

$$dy = dr \sin \theta + r d\theta \cos \theta;$$

$$d^2x = d^2r \cos \theta - 2 dr\, d\theta \sin \theta - r d\theta^2 \cos \theta - r d^2\theta \sin \theta,$$

$$d^2y = d^2r \sin \theta + 2 dr\, d\theta \cos \theta - r d\theta^2 \sin \theta + r d^2\theta \cos \theta.$$

These values substituted in any given formula involving rectangular coordinates, will give the equivalent polar formula in terms of r, θ and their differentials.

The following relations are sometimes useful in dynamical investigations:

$$dx\, \cos\theta + dy\, \sin\theta = dr,$$

$$dy\, \cos\theta - dx\, \sin\theta = r d\theta,$$

$$d^2x \cos\theta + d^2y \sin\theta = d^2r - r d\theta^2,$$

$$d^2y \cos\theta - d^2x \sin\theta = r d^2\theta + 2 dr\, d\theta = \frac{d(r^2 d\theta)}{r}.$$

When θ is taken as the independent variable, $d\theta$ will be constant, and the terms containing $d^2\theta$ will disappear.

(92.) *Rectification.* — Substituting the foregoing values of dx, dy in the equation $ds^2 = dx^2 + dy^2$, we get

$$ds^2 = dr^2 + r^2 d\theta^2,$$

$$\therefore \; ds = \sqrt{(dr^2 + r^2 d\theta^2)},$$

$$\text{and } s = \int d\theta \sqrt{\left(r^2 + \frac{dr^2}{d\theta^2}\right)}.$$

(93.) The value of ds may be immediately deduced from the diagram. Thus if OP and OP' be the radii vectores, subtending the arc $PP' = ds$ and containing the angle $POP' = d\theta$, let Pm be a small arc described with the radius OP and meeting OP' in m; then, when the elements are infinitesimal, this small arc may be regarded as a right line perpendicular to OP'; also, we shall obviously have $mP' = dr$, and $Pm = r d\theta$;

$$\therefore \; ds^2 = PP'^2 = mP'^2 + Pm^2 = dr^2 + r^2 d\theta^2.$$

Several of the subsequent formulæ may also be obtained geometrically from the diagram, and the determination of them in this way would form useful exercises for the student.

(94.) *Perpendicular on the Tangent.*—The perpendicular OH from the origin upon the tangent being denoted by p, we have, art. (73),

$$p = \frac{x \, dy - y \, dx}{ds}.$$

By substituting the preceding polar equivalents, this gives

$$p = \frac{r^2 d\theta}{ds} = \frac{r^2 d\theta}{\sqrt{(dr^2 + r^2 d\theta^2)}}.$$

Cor.— If $u = \dfrac{1}{r}$; then $du = -\dfrac{dr}{r^2}$, and we obtain the neat formula

$$\frac{1}{p^2} = u^2 + \frac{du^2}{d\theta^2}.$$

(95.) *Sectorial Area.*—Conceive two consecutive radii vectores $OP = r$, $OP' = r + dr$ to be drawn, subtending the element $PP' = ds$ of the curve and containing the angle $POP' = d\theta$. The sectorial element thus formed by these radii vectores and ds may be considered as a plane triangle, and the perpendicular from the origin on the opposite side ds produced will obviously be that on the tangent to the curve. Therefore, p denoting this perpendicular, the area of the sectorial element $= \dfrac{pds}{2}$. That is, denoting by S the sectorial area of the curve estimated from a given radius vector, $dS = \dfrac{pds}{2}$. But, art. (94), $p = \dfrac{x\,dy - y\,dx}{ds} = \dfrac{r^2 d\theta}{ds}$,

$$\therefore \; dS = \frac{x\,dy - y\,dx}{2} = \frac{r^2 d\theta}{2}$$

$$\text{and } \; S = \int \frac{r^2 d\theta}{2}.$$

(96.) *Inclination of the tangent with the radius vector.*— Let the angle OPT included by the tangent and radius vector be denoted by P; then by the diagram,

$$\sin P = \frac{OH}{OP} = \frac{p}{r};$$

$$\therefore \; \cos P = \frac{\sqrt{(r^2 - p^2)}}{r}, \qquad \tan P = \frac{p}{\sqrt{(r^2 - p^2)}}.$$

Substituting the value of p, art. (94), these become

$$\sin P = \frac{r\,d\theta}{ds} = \frac{r\,d\theta}{\sqrt{(dr^2 + r^2 d\theta^2)}},$$

$$\cos P = \frac{dr}{ds} = \frac{dr}{\sqrt{(dr^2 + r^2 d\theta^2)}},$$

$$\tan P = \frac{r\,d\theta}{dr}.$$

Cor.—Hence we obtain,

$$ds = \frac{dr}{\cos P} = \frac{r\,dr}{\sqrt{(r^2 - p^2)}},$$

$$d\theta = \frac{dr}{r} \tan P = \frac{p\,dr}{r\sqrt{(r^2 - p^2)}},$$

$$dS = \frac{r^2 d\theta}{2} = \frac{pr\,dr}{2\sqrt{(r^2 - p^2)}}.$$

which are here expressed in terms of the radius vector
and the perpendicular on the tangent.

(97.) *Tangent and Normal.* — Let a
straight line NOT be drawn through
the origin at right angles to the radius
vector OP, and intersecting the tangent
and normal in the points T and N.
This line we shall here designate the
relative axis to the point P. It is
evident that the positions of the tangent and normal with
respect to this axis will enable us to construct them geometrically. The line PT is the *polar tangent,* PN is the *polar
normal,* OT is the *polar subtangent,* and ON is the *polar sub-
normal.* From the angle P, determined in the last article,
the values of these lines are immediately deduced as follows :

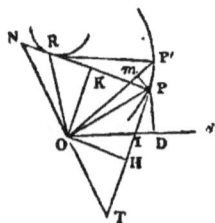

$$PT = \text{polar tangent} = \frac{r}{\cos P} = \frac{r^2}{\sqrt{(r^2-p^2)}} = \frac{r\,ds}{dr},$$

$$PN = \text{polar normal} = \frac{r}{\sin P} = \frac{r^2}{p} = \frac{ds}{d\theta},$$

$$OT = \text{polar subtangent} = r\tan P = \frac{pr}{\sqrt{(r^2-p^2)}} = \frac{r^2 d\theta}{dr},$$

$$ON = \text{polar subnormal} = \frac{r}{\tan P} = \frac{r}{p}\sqrt{(r^2-p^2)} = \frac{dr}{d\theta};$$

$$OH = p = r\sin P = \frac{r^2 d\theta}{ds},$$

$$OK = p_{,} = r\cos P = \sqrt{(r^2-p^2)} = \frac{r\,dr}{ds}.$$

(98.) *Asymptotes.*—If for any finite value of θ the value of r
becomes infinite, the radius vector does not meet the curve
at any finite distance, and therefore it must be parallel to the
tangent which belongs to the corresponding point at the infinite
distance. The polar subtangent $OT = \dfrac{r^2 d\theta}{dr}$ will then become
identical with the perpendicular from the pole on the tangent,
and if its value be finite, the tangent admits of being con-

structed and is then an asymptote to the curve. If the polar subtangent $= 0$, the asymptote passes through the pole and coincides with the radius vector : but if the value of the polar subtangent be infinite, the tangent, being at an infinite distance from the pole, is not an asymptote.

If the diagram be conceived to be turned round into such a position that the radius vector shall proceed from the pole towards the right hand, the rule of signs to be observed in the construction will be simply as follows : If the value of the polar subtangent $OT = \dfrac{r^2 d\theta}{dr}$ be *positive*, it must be measured *downwards*, and if it be *negative*, it must be *measured* upwards; then the right line drawn through the point T parallel to the radius vector, will be the required asymptote.

(99.) A polar curve may have a circular asymptote. If, when the value of the polar angle θ is supposed to proceed positively or negatively to infinity, the point P recedes from the pole until the radius vector ultimately attains, as a superior limit, the finite value a ; then a circle whose centre is the pole O and radius a will evidently be an *exterior asymptotic circle*. But if the point P approaches the pole, until the radius vector reaches as an inferior limit the finite value a, the circle will be an *interior asymptotic circle*.

(100.) *Circle of Curvature.*—The value of the radius of curvature obtained by general differentiation, art. (77), is

$$\rho = \pm \frac{ds^3}{dy\, d^2x - dx\, d^2y}.$$

But, using the polar equivalents, art. (91), we have

$$dy\, d^2x - dx\, d^2y =$$
$$dr(d^2x \sin\theta - d^2y \cos\theta) + r\, d\theta\, (d^2x \cos\theta + d^2y \sin\theta)$$
$$= -dr(2\, dr\, d\theta + r\, d^2\theta) + r\, d\theta\, (d^2r - r\, d\theta^2)$$
$$= -d\theta(r^2 d\theta^2 + 2\, dr^2 - r\, d^2r) - r\, dr\, d^2\theta;$$
$$\therefore\ \rho = \frac{ds^3}{d\theta\, (r^2 d\theta^2 + 2\, dr^2 - r\, d^2r) + r\, dr\, d^2\theta}.$$

By taking θ as the independent variable,

$$\rho = \frac{ds^3}{d\theta(r^2 d\theta^2 + 2\,dr^2 - r\,d^2r)} = \frac{(dr^2 + r^2 d\theta^2)^{\frac{3}{2}}}{d\theta(r^2 d\theta^2 + 2\,dr^2 - r\,d^2r)},$$

which will be positive when the convexity is downwards, and negative when it is upwards.

If $u = \dfrac{1}{r}$; then $r = \dfrac{1}{u}$, $dr = -\dfrac{du}{u^2}$, $d^2r = -\dfrac{d^2u}{u^2} + \dfrac{2\,du^2}{u^3}$,

and the expression for ρ reduces to the convenient form

$$\rho = \frac{\left(u^2 + \dfrac{du^2}{d\theta^2}\right)^{\frac{3}{2}}}{u^3\left(u + \dfrac{d^2u}{d\theta^2}\right)} = \frac{\left(1 + \dfrac{1}{u^2}\dfrac{du^2}{d\theta^2}\right)^{\frac{3}{2}}}{u + \dfrac{d^2u}{d\theta^2}}.$$

(101.) The value of the radius of curvature in terms of r and p may be found as follows:

Referring to the diagram, we have the angle $OPI = P$, $POI = \theta$, and $PID = \omega$; $\therefore \omega = P + \theta$, and $d\omega = dP + d\theta$. But from the values of $\sin P$, $\cos P$, art. (96), we deduce

$$dP = \frac{d\sin P}{\cos P} = \frac{r\,dp - p\,dr}{r\sqrt{(r^2 - p^2)}}.$$

Also, art. (96),

$$ds = \frac{r\,dr}{\sqrt{(r^2 - p^2)}}; \text{ and } d\theta = \frac{p\,dr}{r\sqrt{(r^2 - p^2)}};$$

$$\therefore d\omega = \frac{dp}{\sqrt{(r^2 - p^2)}}.$$

Hence, art. (78),

$$\rho = \frac{ds}{d\omega} = \frac{r\,dr}{dp}.$$

This neat relation may be verified by substituting for dp the differential of the expression $p = \dfrac{r^2 d\theta}{\sqrt{(dr^2 + r^2 d\theta^2)}}$. The result will be found to correspond with the value before obtained.

Examples.

1. In the lemniscate $r^2 = a^2 \cos 2\theta$, $\rho = \dfrac{a^2}{3r}$.

2. In the spiral of Archimedes $r = a\theta$, $\rho = \dfrac{(a^2 + r^2)^{\frac{3}{2}}}{2a^2 + r^2}$.

3. In the reciprocal spiral $u = \dfrac{\theta}{a}$, $\rho = \dfrac{r(a^2 + r^2)^{\frac{3}{2}}}{a^3}$

4. In the cardioid $r = a(1 - \cos\theta)$, $\rho = \dfrac{2}{3}\sqrt{2ar}$.

5. In the logarithmic spiral $p = mr$, $\rho = \dfrac{r}{m}$.

6. In the epicycloid $p^2 = \dfrac{c^2(r^2 - a^2)}{c^2 - a^2}$,

$$\rho = p\frac{c^2 - a^2}{c^2} = \frac{1}{c}\sqrt{(c^2 - a^2)(r^2 - a^2)}.$$

(102.) *Chord of Curvature.*—The portion of the radius vector, produced if necessary, intercepted by the circle of curvature, is called the *chord of curvature*. As this chord evidently subtends an angle, at the centre of the circle, equal to 2P, its value is

$$\text{Chord of Curvature} = 2\rho \sin P = \frac{2p\rho}{r} = \frac{2p\,dr}{dp}.$$

Example 1.—In the lemniscate $r^2 = a^2 \cos 2\theta$, the chord of curvature $= \dfrac{2}{3}r$.

Example 2.—In the cardioid $r = a(1 - \cos\theta)$, the chord of curvature $= \dfrac{4}{3}r$.

(103.) *Evolute and Involute.*—The radius of curvature coincides with the normal and touches the evolute, art. (79). Let $r_{\prime} = OR$, $p_{\prime} = OK$ be the radius vector and perpendicular on the tangent which belong to the *evolute* at the point of contact. By referring to the figure, page 148, it will be seen that p and $p_{\prime}$ constitute a rectangle $HOKP$ with the tangent and normal to the curve; also that $OK^2 = \Pi P^2 = OP^2 - OH^2$ and $OR^2 = RK^2 + OK^2$, that is

$$p_{,}^2 = r^2 - p^2,$$
$$r_{,}^2 = (\rho - p)^2 + p_{,}^2$$
$$= (\rho - p)^2 + r^2 - p^2$$
$$= \rho^2 - 2p\rho + r^2.$$

The value of $\rho = \dfrac{rdr}{dp}$ being previously determined, we can usually by means of these two equations and the equation of the curve $f(r, p) = 0$, eliminate r and p, and so obtain the equation of the evolute in $r_{,}$ and $p_{,}$.

Example 1.—The evolute to the logarithmic spiral $p = mr$ is a similar logarithmic spiral $p_{,} = mr_{,}$.

Example 2.—The evolute to the epicycloid $p^2 = \dfrac{c^2(r^2 - a^2)}{c^2 - a^2}$ is another epicycloid $p_{,}^2 = \dfrac{c^2\left(r_{,}^2 - \dfrac{a^4}{c^2}\right)}{c^2 - a^2}.$

(104.) The value of the radius of curvature may be simply deduced from the equation

$$r_{,}^2 = \rho^2 - 2p\rho + r^2.$$

Since, when we proceed to a consecutive point in the curve, $OR = r_{,}$ and $PR = \rho$, which have reference to the pole O and the intersection R of consecutive normals, do not change, we may differentiate with respect to r and p only, which gives

$$-2\rho dp + 2r dr = 0, \qquad \therefore \rho = \frac{rdr}{dp}.$$

(105.) Let r', p' be the radius vector and perpendicular on the tangent which belong to an *involute* of the curve. As the curve is its evolute, we have from the foregoing equations, substituting $\dfrac{r'dr'}{dp'}$ for ρ',

$$p^2 = r'^2 - p'^2$$
$$r^2 = \left(\frac{r'dr'}{dp'} - p'\right)^2 + r'^2 - p'^2.$$

The values of p and r given by these equations being substituted in the equation of the curve, we shall find an equation involving r', p' and their differentials. If it can be integrated, the equation of the involutes of the curve will thence be found.

(106.) With respect to the evolute, let $\rho_,$ be the radius of curvature at the point R, $ds_,$ the element of the arc, and $\omega_,$ the inclination of the tangent RP with the polar axis. Then

$$\omega_, = \omega + \frac{\pi}{2} \text{ and } ds_, = d\rho;$$

$$\therefore \rho = \frac{ds}{d\omega},$$

$$\rho_, = \frac{ds_,}{d\omega_,} = \frac{d\rho}{d\omega} = \frac{d^2 s}{d\omega^2},$$

the differentiations being with respect to ω as the independent variable.

* These formulæ are useful if s or ρ can be expressed as a function of ω, or when a curve can be reduced to an equation of the form $F(s, \omega) = 0$, or $f(\rho, \omega) = 0$. Thus in the example of the cycloid, page 124, we have

$$\cos \omega = \frac{dx}{ds} = \sqrt{\frac{x}{2a}},$$

$$\rho = 2\sqrt{2a(2a - x)} = 4a \sin \omega;$$

$$\therefore \rho_, = \frac{d\rho}{d\omega} = 4a \cos \omega = -4a \sin \omega_,;$$

and the two equations $\rho = 4a \sin \omega$, and $\rho_, = -4a \sin \omega_,$ which determine the respective curves, show that the evolute to the cycloid is an equal cycloid placed in an inverted position.

(107.) *Positions of Convexity and Points of Inflexion.*— When p is constant or $dp = 0$, the curve becomes a straight

---

* It may here be suggested that a curve may be determined by an equation between any two, or more, of the quantities r, θ, p, ω, ρ, s, and that in particular cases the investigation of the properties of a curve may be greatly simplified by an appropriate selection of variables.

line and therefore has no convexity. On examining the diagram it is evident that if a curve is concave towards the pole, r and p will either both increase or both decrease, and therefore $\dfrac{dp}{dr}$ will be positive; and if the curve is convex towards the pole, r and p will one of them decrease when the other increases, so that $\dfrac{dp}{dr}$ will be negative.

Hence, we have this rule: If

$\dfrac{dp}{dr}$ is $\begin{Bmatrix} \text{positive} \\ \text{negative} \end{Bmatrix}$ the curve is $\begin{Bmatrix} \text{concave} \\ \text{convex} \end{Bmatrix}$ towards the pole.

When $\dfrac{dp}{dr}$ changes sign by passing through 0 or $\dfrac{1}{0}$ the direction of curvature will become reversed, and this will indicate a point of *inflexion*.

(108.) *Locus of the point where the perpendicular meets the tangent.*—Let it be required to find the equation to the curve which is the locus of the point H, where the perpendicular from the pole intersects the tangent. Denote the radius vector OH of this curve by $r_{\prime\prime}$, and the corresponding polar angle and perpendicular upon the tangent by $\theta_{\prime\prime}$ and $p_{\prime\prime}$. Then we shall have $p = r_{\prime\prime}$, and, since O H is perpendicular to P H, the angle between two consecutive positions of O H will be equal to that between corresponding positions of the tangent PH; that is, $d\theta_{\prime\prime} = d\omega$. But, art. (101),

$$d\omega = \frac{dp}{\sqrt{(r^2 - p^2)}} = \frac{dr_{\prime\prime}}{\sqrt{(r^2 - r_{\prime\prime}^2)}},$$

and, art. (96), $d\theta_{\prime\prime} = \dfrac{p_{\prime\prime}\, dr_{\prime\prime}}{r_{\prime\prime}\sqrt{(r_{\prime\prime}^2 - p_{\prime\prime}^2)}};$

$\therefore \dfrac{1}{\sqrt{(r^2 - r_{\prime\prime}^2)}} = \dfrac{p_{\prime\prime}}{r_{\prime\prime}\sqrt{(r_{\prime\prime}^2 - p_{\prime\prime}^2)}},$ and $r = \dfrac{r_{\prime\prime}^2}{p_{\prime\prime}}.$

Hence, if the polar equation to the given curve be $f(p, r) = 0$, that of the locus of H will be $f\left(r_{\prime\prime}, \dfrac{r_{\prime\prime}^2}{p_{\prime\prime}}\right) = 0$, being obtained by simply substituting the values $p = r_{\prime\prime}$, $r = \dfrac{r_{\prime\prime}^2}{p_{\prime\prime}}$ in the given equation.

Example 1.—In the case of the logarithmic spiral, the locus of the point H is an equal and similar logarithmic spiral.

Example 2.—In the case of the rectangular hyperbola, the locus is a lemniscate.

The preceding articles present a complete digest of the most useful formulæ which relate to curves referred to polar coordinates, and by them we are enabled to trace and discuss all the peculiarities and properties of curves from their polar equations.

(109.) For convenience of reference, we shall here collect together the equations of the principal known curves; and we shall then conclude with some general theorems, which have been deferred for insertion at the end of the volume.

1. The *Parabola;* referred to its vertex and axis, $y^2 = 4mx$; the focus being the pole, the polar equation is $r = \dfrac{2m}{1 + \cos\theta}$, or $p^2 = mr$.

2. The *Ellipse;* referred to its centre and principal axes, the equation is $\dfrac{x^2}{a^2} + \dfrac{y^2}{b^2} = 1$; when the centre is the pole, the polar equation is $r^2 = a^2 \left(\dfrac{1-e^2}{1 - e^2 \cos^2\theta} \right)$; and, when the focus is the pole, it is $r = \dfrac{a(1-e^2)}{1 + e\cos\theta}$, or $p = b \sqrt{\dfrac{r}{2a-r}}$, where $e = \dfrac{\sqrt{(a^2 - b^2)}}{a}$.

3. The *Hyperbola.* — Referred to its centre and principal axes, the equation is $\dfrac{x^2}{a^2} - \dfrac{y^2}{b^2} = 1$; when the centre is the pole, $r^2 = a^2 \dfrac{e^2 - 1}{1 - e^2 \cos^2\theta}$; and when the focus is the pole, $r = \dfrac{a(e^2 - 1)}{1 + e\cos\theta}$ or $p = b \sqrt{\dfrac{r}{2a + r}}$, where $e = \dfrac{\sqrt{(a^2 + b^2)}}{a}$. The hyperbola has two asymptotes.

4. The *Equilateral Hyperbola,* when referred to its asymptotes, has for its equation $2xy = a^2$; and the polar equation is $r^2 = \dfrac{a^2}{\sin 2\theta}$, or $p = \dfrac{a^2}{r}$.

5. The *Cycloid.*—Referred to its vertex and axis, the equation is

$$y = \sqrt{(2ax - x^2)} + a \operatorname{vers}^{-1}\frac{x}{a},$$

which may be otherwise stated $x = a(1 - \cos\varphi)$, $y = a(\varphi + \sin\varphi)$.

6. The *Catenary.* — Referred to a point at the distance c below the lowest point of the curve, with the axis of x horizontal ; its equation is

$$y = \frac{c}{2}\left(e^{\frac{x}{c}} + e^{-\frac{x}{c}}\right);$$ and the radius of curvature $\rho = -\frac{y^2}{c}$ is equal to the

normal, but drawn in the opposite direction.

7. The *Logarithmic Curve.*—Its equation is $y = ce^{\frac{x}{a}}$; the subtangent $= a$ is constant, and the negative axis of x is an asymptote.

8. The *Cissoid* of Diocles.— Its equation is $y^2 = \dfrac{x^3}{2a - x}$; the origin is a cusp of the first kind, and the curve has evidently an asymptote perpendicular to the axis of x at the distance $x = 2a$.

9. The *Conchoid* of Nicomedes.—Its equation is $x^2 y^2 = (a^2 - y^2)(b + y)^2$; the axis of y contains a double point, and the axis of x is an asymptote.

10. The *Lemniscate* of Bernoulli. — Its form resembles the symbol ∞ , and, referred to its centre or double point, the equation is

$$(x^2 + y^2)^2 = a^2(x^2 - y^2); \text{ or } r^2 = a^2 \cos 2\theta, \text{ or } p = \frac{r^3}{a^2}.$$

11. The *Witch* of Agnesi. — Referred to its vertex, the equation is $y^2 = \dfrac{a^2 x}{a - x}$; it has inflexions at the points $x = \dfrac{a}{4}, y = \pm \dfrac{a}{\sqrt{3}}$, and an asymptote perpendicular to the axis at the distance $x = a$.

12. The *Spiral* of Archimedes.— The polar equation is

$$r = a\theta, \text{ or } p = \frac{r^2}{\sqrt{(a^2 + r^2)}}.$$

13. The *Reciprocal Spiral.*—Its polar equation is

$$r = \frac{a}{\theta}, \text{ or } p = \frac{ar}{\sqrt{(a^2 + r^2)}}.$$

14. The *Logarithmic Spiral.*—Its polar equation is $r = a^\theta$; or $p = mr$; the curve intersects its radius vector at a constant angle P ; and its evolute and involute are spirals equal to the original one.

15. The *Cardioid.*—Its polar equation is $r = a(1 - \cos\theta)$ or $r^3 = 2ap^2$; the origin is a cusp of the first kind, and its evolute is another cardioid ; also the lines drawn through the pole, and intercepted by the curve, are all of the same length $2a$.

16. *Quadratrix* of Dinostratus.— Its equation is $y = x \tan \dfrac{\pi(a - x)}{2a}$,

and it has an infinite number of asymptotes perpendicular to the axis of x. When $x = 0$, $y = 0 \times \infty = \dfrac{2a}{\pi}$.

17. *Quadratrix* of Tschirnhausen.—Its equation is $y = a \sin \dfrac{\pi x}{2a}$, and it has inflexions at the points where $y = 0$.

18. *Companion to the Cycloid:* $x = a(1 - \cos\phi)$, $y = a\phi$.

19. *Trochoid;* $x = a(1 - n \cos\phi)$, $y = a(\phi - n \sin\phi)$.

20. *Epitrochoid;* $x = (a + b) \cos\phi - h \cos\left(\dfrac{a + b}{b}\right)\phi$,

$$y = (a + b) \sin\phi - h \sin\left(\dfrac{a + b}{b}\right)\phi.$$

21. When $h = b$, this becomes the *Epicycloid;* and when also $a = b$, it becomes the *Cardioid.*

22. *Hypotrochoid;* $x = (a - b) \cos\phi + h \cos\left(\dfrac{a - b}{b}\right)\phi$,

$$y = (a - b) \sin\phi - h \sin\left(\dfrac{a - b}{b}\right)\phi.$$

23. When $h = b$, this becomes the *Hypocycloid;* when $b = \dfrac{a}{4}$, it gives $x^{\frac{2}{3}} + y^{\frac{2}{3}} = a^{\frac{2}{3}}$; and when $b = \dfrac{a}{2}$, it becomes an *Ellipse.*

24. The *Lituus.*—Its polar equation is $r^2 = \dfrac{a^2}{\theta}$.

Euler's Theorems on Homogeneous Functions.

(110.) If $u = f(x, y, z, \&c.)$ be a homogeneous function of n dimensions and of any number of variables; then

$$x\left(\frac{du}{dx}\right) + y\left(\frac{du}{dy}\right) + z\left(\frac{du}{dz}\right) + \&c. \qquad = nu,$$

$$x^2\left(\frac{d^2u}{dx^2}\right) + y^2\left(\frac{d^2u}{dy^2}\right) + \ldots\ldots + 2xy\left(\frac{d^2u}{dx\,dy}\right) + \&c.$$
$$= n(n-1)\,u,$$

$$x^3\left(\frac{d^3u}{dx^3}\right) + y^3\left(\frac{d^3u}{dy^3}\right) + \ldots\ldots + 3x^2y\left(\frac{d^3u}{dx^2\,dy}\right) + \&c.$$
$$= n(n-1)(n-2)\,u,$$

$$\&c. \qquad\qquad \&c. \qquad\qquad \&c.$$

Since the function u is homogeneous and n is the sum of the exponents of the variables in each term, if for x, y, z, &c. there be substituted $(1 + a)x$, $(1 + a)y$, $(1 + a)z$, &c. it is evident that the value of u will become $(1 + a)^n u$; that is

$$(1 + a)^n u = f(x + ax, y + ay, z + az, \&c.)$$

The first of these being expanded by the binomial theorem, and the second by the formula of art. (47), by equating the coefficients of the like powers of the arbitrary quantity a, we obtain the elegant relations stated in the theorems.

Laplace's Theorem.

(111.) If $y = f(z + x\phi y)$, in which y is an implicit function of two variables x and z depending on the forms of the functions characterized by f and ϕ; then the development of any other function Fy may be obtained from the following general theorem :

$$Fy = Ffz + \frac{d \cdot Ffz}{dz} (\phi fz) \frac{x}{1} + \frac{d}{dz} \left\{ \frac{d \cdot Ffz}{dz} (\phi fz)^2 \right\} \frac{x^2}{1.2}$$

$$+ \frac{d^2}{dz^2} \left\{ \frac{d \cdot Ffz}{dz} (\phi fz)^3 \right\} \frac{x^3}{1.2.3} \cdots \cdots$$

$$+ \frac{d^{n-1}}{dz^{n-1}} \left\{ \frac{d \cdot Ffz}{dz} (\phi fz)^n \right\} \frac{x^n}{1.2 \ldots n} + \&c.$$

By considering $u = Fy$ as a function of x its expansion in powers of x, art. (46), is

$$u = u_0 + \frac{x}{1} \left(\frac{du}{dx} \right)_0 + \frac{x^2}{1.2} \left(\frac{d^2 u}{dx^2} \right)_0 + \frac{x^3}{1.2.3} \left(\frac{d^3 u}{dx^3} \right)_0 + \&c. \ldots (a)$$

where the values of u_0 and the differential coefficients, as indicated, are to be taken when $x = 0$. For the investigation of the proposed theorem it will therefore only be requisite to determine the values of these coefficients. Let

$$\beta = z + x\phi y;$$

then $y = f\beta = f(z + x\phi y)$. By differentiating first with respect to x and then with respect to z, we have

$$\frac{dy}{dx} = \left(\phi y + x\phi' y \frac{dy}{dx}\right) f'\beta, \qquad \text{or} \qquad \frac{dy}{dx} = \frac{\phi y f'\beta}{1 - x\phi' y f'\beta};$$

$$\frac{dy}{dz} = \left(1 + x\phi' y \frac{dy}{dz}\right) f'\beta, \qquad \text{or} \qquad \frac{dy}{dz} = \frac{f'\beta}{1 - x\phi' y f'\beta};$$

$$\therefore \frac{dy}{dx} = \frac{dy}{dz} \phi y.$$

This equation being independent of the form of the function $y = f\beta$ must evidently be true if y be replaced by any function of β or by any function of y. Substituting therefore $u = Fy$, we get

$$\frac{du}{dx} = \frac{du}{dz} \phi y \dots \dots (1).$$

Again, since u is a function of y, which is a function of two variables x and z, we have, art. (37) and this equation (1),

$$\frac{d^2 u}{dx^2} = \frac{d}{dx} \frac{du}{dz} \phi y = \frac{d}{dz} \frac{du}{dx} \phi y = \frac{d}{dz} \left\{ \frac{du}{dz} (\phi y)^2 \right\} \dots (2),$$

$$\frac{d^3 u}{dx^3} = \frac{d}{dx} \frac{d}{dz} \frac{du (\phi y)^2}{dz} = \frac{d^2}{dz^2} \frac{du (\phi y)^2}{dx} = \frac{d^2}{dz^2} \left\{ \frac{du}{dz} (\phi y)^3 \right\}$$
$$\dots (3),$$

$$\text{\&c.} \qquad \text{\&c.} \qquad \text{\&c.}$$

$$\frac{d^n u}{dx^n} = \frac{d}{dx} \frac{d^{n-2}}{dz^{n-2}} \frac{du (\phi y)^{n-1}}{dz} = \frac{d^{n-1}}{dz^{n-1}} \frac{du (\phi y)^{n-1}}{dx}$$
$$= \frac{d^{n-1}}{dz^{n-1}} \left\{ \frac{du}{dz} (\phi y)^n \right\} \dots (n),$$

In deducing the values of the differential coefficients when $x = 0$ we may obviously make $x = 0$ before differentiating; that is, we may at once use $u_0 = Fy_0 = Ffz$, and $\phi y_0 = \phi fz$. Thus we find,

$$u_0 = \mathrm{F}fz,$$

$$\left(\frac{du}{dx}\right)_0 = \frac{d \cdot \mathrm{F}fz}{dz}\,(\phi fz),$$

$$\left(\frac{d^2u}{dx^2}\right)_0 = \frac{d}{dz}\left\{\frac{d \cdot \mathrm{F}fz}{dz}\,(\phi fz)^2\right\},$$

$$\left(\frac{d^3u}{dx^3}\right)_0 = \frac{d^2}{dz^2}\left\{\frac{d \cdot \mathrm{F}fz}{dz}\,(\phi fz)^3\right\},$$

&c.   &c.

$$\left(\frac{d^nu}{dx^n}\right)_0 = \frac{d^{n-1}}{dz^{n-1}}\left\{\frac{d \cdot \mathrm{F}fz}{dz}\,(\phi fz)^n\right\};$$

and by substituting these values in

$$u = u_0 + \frac{x}{1}\left(\frac{du}{dx}\right)_0 + \frac{x^2}{1.2}\left(\frac{d^2u}{dx^2}\right)_0 + \frac{x^3}{1.2.3}\left(\frac{d^3u}{dx^3}\right)_0 + \&c.$$

we obtain the theorem stated.

Lagrange's Theorem.

(112.) If $y = z + x\phi y$, where ϕy denotes a given function; then the development of another function $\mathrm{F}y$ in ascending powers of x will be

$$\mathrm{F}y = \mathrm{F}z + \frac{d \cdot \mathrm{F}z}{dz}\,(\phi z)\frac{x}{1} + \frac{d}{dz}\left\{\frac{d \cdot \mathrm{F}z}{dz}\,(\phi z)^2\right\}\frac{x^2}{1.2}$$

$$+ \frac{d^2}{dz^2}\left\{\frac{d \cdot \mathrm{F}z}{dz}\,(\phi z)^3\right\}\frac{x^3}{1.2.3}$$

$$\cdots\cdots\cdots\cdots + \frac{d^{n-1}}{dz^{n-1}}\left\{\frac{d \cdot \mathrm{F}z}{dz}\,(\phi z)^n\right\}\frac{x^n}{1.2\cdots n} + \&c.$$

This is a case of Laplace's more general theorem, from which it immediately follows on making $fz = z$; and when $\phi z = 1$, it becomes Taylor's theorem.

INDEX.

H

THE END.

PRINTED BY VIRTUE AND CO., CITY ROAD, LONDON.

𝔚eale's 𝔕udimentary 𝔖eries

PHILADELPHIA, 1876.
THE PRIZE MEDAL
Was awarded to the Publishers for
Books: Rudimentary Scientific,
"WEALE'S SERIES," ETC.

A NEW LIST OF
WEALE'S SERIES
RUDIMENTARY SCIENTIFIC, EDUCATIONAL, AND CLASSICAL.

LONDON, 1862.
THE PRIZE MEDAL
Was awarded to the Publishers of
"WEALE'S SERIES."

These popular and cheap Series of Books, now comprising nearly Three Hundred distinct works in almost every department of Science, Art, and Education, are recommended to the notice of Engineers, Architects, Builders, Artisans, and Students generally, as well as to those interested in Workmen's Libraries, Free Libraries, Literary and Scientific Institutions, Colleges, Schools, Science Classes, &c., &c.

N.B.—In ordering from this List it is recommended, as a means of facilitating business and obviating error, to quote the numbers affixed to the volumes, as well as the titles and prices.
*** The books are bound in limp cloth, unless otherwise stated.

RUDIMENTARY SCIENTIFIC SERIES.

ARCHITECTURE, BUILDING, ETC.

No.
16. *ARCHITECTURE—ORDERS*—The Orders and their Æsthetic Principles. By W. H. LEEDS. Illustrated. 1s. 6d.
17. *ARCHITECTURE—STYLES*—The History and Description of the Styles of Architecture of Various Countries, from the Earliest to the Present Period. By T. TALBOT BURY, F.R.I.B.A., &c. Illustrated. 2s.
*** ORDERS AND STYLES OF ARCHITECTURE, in One Vol., 3s. 6d.
18. *ARCHITECTURE—DESIGN*—The Principles of Design in Architecture, as deducible from Nature and exemplified in the Works of the Greek and Gothic Architects. By E. L. GARBETT, Architect. Illustrated. 2s.
*** *The three preceding Works, in One handsome Vol., half bound, entitled* "MODERN ARCHITECTURE," *Price 6s.*
22. *THE ART OF BUILDING*, Rudiments of. General Principles of Construction, Materials used in Building, Strength and Use of Materials, Working Drawings, Specifications, and Estimates. By EDWARD DOBSON, M.R.I.B.A., &c. Illustrated. 2s.
23. *BRICKS AND TILES*, Rudimentary Treatise on the Manufacture of; containing an Outline of the Principles of Brickmaking. By EDW. DOBSON, M.R.I.B.A. With Additions by C. TOMLINSON, F.R.S. Illustrated. 3s.

Architecture, Building, etc., *continued.*

25. *MASONRY AND STONECUTTING,* Rudimentary Treatise on ; in which the Principles of Masonic Projection and their application to the Construction of Curved Wing-Walls, Domes, Oblique Bridges, and Roman and Gothic Vaulting, are concisely explained. By EDWARD DOBSON, M.R.I.B.A., &c. Illustrated with Plates and Diagrams. 2s. 6d.

44. *FOUNDATIONS AND CONCRETE WORKS,* a Rudimentary Treatise on : containing a Synopsis of the principal cases of Foundation Works, with the usual Modes of Treatment, and Practical Remarks on Footings, Planking, Sand, Concrete, Béton, Pile-driving, Caissons, and Cofferdams. By E. DOBSON, M.R.I.B.A., &c. Fourth Edition, revised by GEORGE DODD, C.E. Illustrated. 1s. 6d.

42. *COTTAGE BUILDING.* By C. BRUCE ALLEN, Architect. Eleventh Edition, revised and enlarged. Numerous Illustrations. 1s. 6d.

45. *LIMES, CEMENTS, MORTARS, CONCRETES, MASTICS,* PLASTERING, &c. By G. R. BURNELL, C.E. Ninth Edition. 1s. 6d.

57. *WARMING AND VENTILATION,* a Rudimentary Treatise on ; being a concise Exposition of the General Principles of the Art of Warming and Ventilating Domestic and Public Buildings, Mines, Lighthouses, Ships, &c. By CHARLES TOMLINSON, F.R.S., &c. Illustrated. 3s.

83**. *CONSTRUCTION OF DOOR LOCKS.* Compiled from the Papers of A. C. HOBBS, Esq., of New York, and Edited by CHARLES TOMLINSON, F.R.S. To which is added, a Description of Fenby's Patent Locks, and a Note upon IRON SAFES by ROBERT MALLET, M.I.C.E. Illus. 2s. 6d.

111. *ARCHES, PIERS, BUTTRESSES, &c.:* Experimental Essays on the Principles of Construction in ; made with a view to their being useful to the Practical Builder. By WILLIAM BLAND. Illustrated. 1s. 6d.

116. *THE ACOUSTICS OF PUBLIC BUILDINGS;* or, The Principles of the Science of Sound applied to the purposes of the Architect and Builder. By T. ROGER SMITH, M.R.I.B.A., Architect. Illustrated. 1s. 6d.

124. *CONSTRUCTION OF ROOFS,* Treatise on the, as regards Carpentry and Joinery. Deduced from the Works of ROBISON, PRICE, and TREDGOLD. Illustrated. 1s. 6d.

127. *ARCHITECTURAL MODELLING IN PAPER,* the Art of. By T. A. RICHARDSON, Architect. Illustrated. 1s. 6d.

128. *VITRUVIUS—THE ARCHITECTURE OF MARCUS VITRUVIUS POLLO.* In Ten Books. Translated from the Latin by JOSEPH GWILT, F.S.A., F.R.A.S. With 23 Plates. 5s.

130. *GRECIAN ARCHITECTURE,* An Inquiry into the Principles of Beauty in ; with a Historical View of the Rise and Progress of the Art in Greece. By the EARL OF ABERDEEN. 1s.
⁎⁎⁎ *The two Preceding Works in One handsome Vol., half bound, entitled* "ANCIENT ARCHITECTURE." *Price 6s.*

132. *DWELLING-HOUSES,* a Rudimentary Treatise on the Erection of. By S. H. BROOKS, Architect. New Edition, with Plates. 2s. 6d.

156. *QUANTITIES AND MEASUREMENTS,* How to Calculate and Take them in Bricklayers', Masons', Plasterers', Plumbers', Painters', Paperhangers', Gilders', Smiths', Carpenters', and Joiners' Work. By A. C. BEATON, Architect and Surveyor. New and Enlarged Edition. Illus. 1s. 6d.

175. *LOCKWOOD & CO.'S BUILDER'S AND CONTRACTOR'S* PRICE BOOK, for 1877, containing the latest Prices of all kinds of Builders' Materials and Labour, and of all Trades connected with Building : Lists of the Members of the Metropolitan Board of Works, of Districts, District Officers, and District Surveyors, and the Metropolitan Bye-laws. Edited by FRANCIS T. W. MILLER, Architect and Surveyor. 3s. 6d.

182. *CARPENTRY AND JOINERY*—THE ELEMENTARY PRINCIPLES OF CARPENTRY. Chiefly composed from the Standard Work of THOMAS TREDGOLD, C.E. With Additions from the Works of the most Recent Authorities, and a TREATISE ON JOINERY by E. WYNDHAM TARN, M.A. Numerous Illustrations. 3s. 6d.

Architecture, Building, etc., *continued.*

182*. *CARPENTRY AND JOINERY. ATLAS* of 35 Plates to accompany the foregoing book. With Descriptive Letterpress. 4to. 6s.

187. *HINTS TO YOUNG ARCHITECTS.* By GEORGE WIGHTWICK. New, Revised, and enlarged Edition. By G. HUSKISSON GUILLAUME, Architect. With numerous Woodcuts. 3s. 6d.

188. *HOUSE PAINTING, GRAINING, MARBLING, AND SIGN WRITING:* A Practical Manual of. With 9 Coloured Plates of Woods and Marbles, and nearly 150 Wood Engravings. By ELLIS A. DAVIDSON. Second Edition, carefully revised, 5s. *[Just published.*

189. *THE RUDIMENTS OF PRACTICAL BRICKLAYING.* In Six Sections: General Principles; Arch Drawing, Cutting, and Setting; Pointing; Paving, Tiling, Materials; Slating and Plastering; Practical Geometry, Mensuration, &c. By ADAM HAMMOND. Illustrated. 1s. 6d.

191. *PLUMBING.* A Text-Book to the Practice of the Art or Craft of the Plumber. With Chapters upon House Drainage, embodying the latest Improvements. Containing about 300 Illustrations. By W. P. BUCHAN, Sanitary Engineer. 3s. *[Just published.*

192. *THE TIMBER IMPORTER'S, TIMBER MERCHANT'S,* and BUILDER'S STANDARD GUIDE; comprising copious and valuable Memoranda for the Retailer and Builder. By RICHARD E. GRANDY. Second Edition, Revised. 3s.

CIVIL ENGINEERING, ETC.

13. *CIVIL ENGINEERING,* the Rudiments of; for the Use of Beginners, for Practical Engineers, and for the Army and Navy. By HENRY LAW, C.E. Including a Section on Hydraulic Engineering, by GEORGE R. BURNELL, C.E. 5th Edition, with Notes and Illustrations by ROBERT MALLET, A.M., F.R.S. Illustrated with Plates and Diagrams. 5s.

29. *THE DRAINAGE OF DISTRICTS AND LANDS.* By G. DRYSDALE DEMPSEY, C.E. New Edition, enlarged. Illustrated. 1s. 6d.

30. *THE DRAINAGE OF TOWNS AND BUILDINGS.* By G. DRYSDALE DEMPSEY, C.E. New Edition. Illustrated. 2s. 6d.
⁎ With "*Drainage of Districts and Lands,*" in One Vol., 3s. 6d.

31. *WELL-DIGGING, BORING, AND PUMP-WORK.* By JOHN GEORGE SWINDELL, Assoc. R.I.B.A. New Edition, revised by G. R. BURNELL, C.E. Illustrated. 1s. 6d.

35. *THE BLASTING AND QUARRYING OF STONE,* for Building and other Purposes. With Remarks on the Blowing up of Bridges. By Gen. Sir JOHN BURGOYNE, Bart., K.C.B. Illustrated. 1s. 6d.

43. *TUBULAR AND OTHER IRON GIRDER BRIDGES.* Particularly describing the BRITANNIA and CONWAY TUBULAR BRIDGES. With a Sketch of Iron Bridges, and Illustrations of the Application of Malleable Iron to the Art of Bridge Building. By G. D. DEMPSEY, C.E. New Edition, with Illustrations. 1s. 6d.

62. *RAILWAY CONSTRUCTION,* Elementary and Practical Instruction on. By Sir MACDONALD STEPHENSON, C.E. New Edition, enlarged by EDWARD NUGENT, C.E. Plates and numerous Woodcuts. 3s.

80*. *EMBANKING LANDS FROM THE SEA,* the Practice of. Treated as a Means of Profitable Employment for Capital. With Examples and Particulars of actual Embankments, and also Practical Remarks on the Repair of old Sea Walls. By JOHN WIGGINS, F.G.S. New Edition, with Notes by ROBERT MALLET, F.R.S. 2s.

81. *WATER WORKS,* for the Supply of Cities and Towns. With a Description of the Principal Geological Formations of England as influencing Supplies of Water; and Details of Engines and Pumping Machinery for raising Water. By SAMUEL HUGHES, F.G.S., C.E. New Edition, revised and enlarged, with numerous Illustrations. 4s.

82**. *GAS WORKS,* and the Practice of Manufacturing and Distributing Coal Gas. By SAMUEL HUGHES, C.E. New Edition, revised by W. RICHARDS, C.E. Illustrated. 3s. 6d.

Civil Engineering, etc., *continued.*

117. *SUBTERRANEOUS SURVEYING;* an Elementary and Practical Treatise on. By THOMAS FENWICK. Also the Method of Conducting Subterraneous Surveys without the Use of the Magnetic Needle, and other modern Improvements. By THOMAS BAKER, C.E. Illustrated. 2s. 6d.

118. *CIVIL ENGINEERING IN NORTH AMERICA,* a Sketch of. By DAVID STEVENSON, F.R.S.E., &c. Plates and Diagrams. 3s.

121. *RIVERS AND TORRENTS.* With the Method of Regulating their Courses and Channels. By Professor PAUL FRISI, F.R.S., of Milan. To which is added, AN ESSAY ON NAVIGABLE CANALS. Translated by Major-General JOHN GARSTIN, of the Bengal Engineers. Plates. 2s. 6d.

MECHANICAL ENGINEERING, ETC.

33. *CRANES,* the Construction of, and other Machinery for Raising Heavy Bodies for the Erection of Buildings, and for Hoisting Goods. By JOSEPH GLYNN, F.R.S., &c. Illustrated. 1s. 6d.

34. *THE STEAM ENGINE,* a Rudimentary Treatise on. By Dr. LARDNER. Illustrated. 1s. 6d.

59. *STEAM BOILERS:* their Construction and Management. By R. ARMSTRONG, C.E. Illustrated. 1s. 6d.

63. *AGRICULTURAL ENGINEERING:* Farm Buildings, Motive Power, Field Machines, Machinery, and Implements. By G. H. ANDREWS, C.E. Illustrated. 3s.

67. *CLOCKS, WATCHES, AND BELLS,* a Rudimentary Treatise on. By Sir EDMUND BECKETT (late EDMUND BECKETT DENISON, LL.D., Q.C.). A new, Revised, and considerably Enlarged Edition (the 6th), with very numerous Illustrations. 4s. 6d. [*Just published.*

77*. *THE ECONOMY OF FUEL,* particularly with Reference to Reverbatory Furnaces for the Manufacture of Iron, and to Steam Boilers. By T. SYMES PRIDEAUX. 1s. 6d.

82. *THE POWER OF WATER,* as applied to drive Flour Mills, and to give motion to Turbines and other Hydrostatic Engines. By JOSEPH GLYNN, F.R.S., &c. New Edition, Illustrated. 2s.

98. *PRACTICAL MECHANISM,* the Elements of; and Machine Tools. By T. BAKER, C.E. With Remarks on Tools and Machinery, by J. NASMYTH, C.E. Plates. 2s. 6d.

114. *MACHINERY,* Elementary Principles of, in its Construction and Working. Illustrated by numerous Examples of Modern Machinery for different Branches of Manufacture. By C. D. ABEL, C.E. 1s. 6d.

15. *ATLAS OF PLATES.* Illustrating the above Treatise. By C. D. ABEL, C.E. 7s. 6d.

125. *THE COMBUSTION OF COAL AND THE PREVENTION* OF SMOKE, Chemically and Practically Considered. With an Appendix. By C. WYE WILLIAMS, A.I.C.E. Plates. 3s.

139. *THE STEAM ENGINE,* a Treatise on the Mathematical Theory of, with Rules at length, and Examples for the Use of Practical Men. By T. BAKER, C.E. Illustrated. 1s. 6d.

162. *THE BRASS FOUNDER'S MANUAL;* Instructions for Modelling, Pattern-Making, Moulding, Turning, Filing, Burnishing, Bronzing, &c. With copious Receipts, numerous Tables, and Notes on Prime Costs and Estimates. By WALTER GRAHAM. Illustrated. 2s. 6d.

164. *MODERN WORKSHOP PRACTICE,* as applied to Marine, Land, and Locomotive Engines, Floating Docks, Dredging Macnines, Bridges, Cranes, Ship-building, &c., &c. By J. G. WINTON. Illustrated. 3s.

165. *IRON AND HEAT,* exhibiting the Principles concerned in the Construction of Iron Beams, Pillars, and Bridge Girders, and the Action of Heat in the Smelting Furnace. By J. ARMOUR, C.E. 2s. 6d.

Mechanical Engineering, etc., *continued.*

166. *POWER IN MOTION:* Horse-Power, Motion, Toothed-Wheel Gearing, Long and Short Driving Bands, Angular Forces. By JAMES ARMOUR, C.E. With 73 Diagrams. 2s. 6d.

167. *THE APPLICATION OF IRON TO THE CONSTRUCTION* OF BRIDGES, GIRDERS, ROOFS, AND OTHER WORKS. Showing the Principles upon which such Structures are designed, and their Practical Application. By FRANCIS CAMPIN, C.E. Second Edition, revised and corrected. Numerous Woodcuts. 2s. 6d.

171. *THE WORKMAN'S MANUAL OF ENGINEERING* DRAWING. By JOHN MAXTON, Engineer, Instructor in Engineering Drawing, Royal Naval College, Greenwich, formerly of R.S.N.A., South Kensington. Third Edition. Illustrated with 7 Plates and nearly 350 Woodcuts. 3s. 6d.

190. *STEAM AND THE STEAM ENGINE*, Stationary and Portable. An elementary treatise on. Being an extension of Mr. John Sewell's "Treatise on Steam." By D. KINNEAR CLARK, C.E., M.I.C.E. Author of "Railway Machinery," "Railway Locomotives," &c., &c. With numerous Illustrations. 3s. 6d. *[Just published.*

SHIPBUILDING, NAVIGATION, MARINE ENGINEERING, ETC.

51. *NAVAL ARCHITECTURE*, the Rudiments of; or, an Exposition of the Elementary Principles of the Science, and their Practical Application to Naval Construction. Compiled for the Use of Beginners. By JAMES PEAKE, School of Naval Architecture, H.M. Dockyard, Portsmouth. Fourth Edition, corrected, with Plates and Diagrams. 3s. 6d.

53*. *SHIPS FOR OCEAN AND RIVER SERVICE*, Elementary and Practical Principles of the Construction of. By HAKON A. SOMMERFELDT, Surveyor of the Royal Norwegian Navy. With an Appendix. 1s.

53**. *AN ATLAS OF ENGRAVINGS* to Illustrate the above. Twelve large folding plates. Royal 4to, cloth. 7s. 6d.

54. *MASTING, MAST-MAKING, AND RIGGING OF SHIPS*, Rudimentary Treatise on. Also Tables of Spars, Rigging, Blocks; Chain, Wire, and Hemp Ropes, &c., relative to every class of vessels. Together with an Appendix of Dimensions of Masts and Yards of the Royal Navy of Great Britain and Ireland. By ROBERT KIPPING, N.A. Fourteenth Edition. Illustrated. 2s.

54*. *IRON SHIP-BUILDING.* With Practical Examples and Details for the Use of Ship Owners and Ship Builders. By JOHN GRANTHAM, Consulting Engineer and Naval Architect. Fifth Edition, with important Additions. 4s.

54**. *AN ATLAS OF FORTY PLATES* to Illustrate the above. Fifth Edition. Including the latest Examples, such as H.M. Steam Frigates "Warrior," "Hercules," "Bellerophon;" H.M. Troop Ship "Serapis," Iron Floating Dock, &c., &c. 4to, boards. 38s.

55. *THE SAILOR'S SEA BOOK:* a Rudimentary Treatise on Navigation. I. How to Keep the Log and Work it off. II. On Finding the Latitude and Longitude. By JAMES GREENWOOD, B.A., of Jesus College, Cambridge. To which are added, Directions for Great Circle Sailing; an Essay on the Law of Storms and Variable Winds; and Explanations of Terms used in Ship-building. Ninth Edition, with several Engravings and Coloured Illustrations of the Flags of Maritime Nations. 2s.

80. *MARINE ENGINES, AND STEAM VESSELS*, a Treatise on. Together with Practical Remarks on the Screw and Propelling Power, as used in the Royal and Merchant Navy. By ROBERT MURRAY, C.E., Engineer-Surveyor to the Board of Trade. With a Glossary of Technical Terms, and their Equivalents in French, German, and Spanish. Fifth Edition, revised and enlarged. Illustrated. 3s.

Shipbuilding, Navigation, etc., *continued.*

83*bis.* *THE FORMS OF SHIPS AND BOATS:* Hints, Experimentally Derived, on some of the Principles regulating Ship-building. By W. BLAND. Sixth Edition, revised, with numerous Illustrations and Models. 1s. 6d.

99. *NAVIGATION AND NAUTICAL ASTRONOMY*, in Theory and Practice. With Attempts to facilitate the Finding of the Time and the Longitude at Sea. By J. R. YOUNG, formerly Professor of Mathematics in Belfast College. Illustrated. 2s. 6d.

100*. *TABLES* intended to facilitate the Operations of Navigation and Nautical Astronomy, as an Accompaniment to the above Book. By J. R. YOUNG. 1s. 6d.

106. *SHIPS' ANCHORS*, a Treatise on. By GEORGE COTSELL, N.A. Illustrated. 1s. 6d.

149. *SAILS AND SAIL-MAKING*, an Elementary Treatise on. With Draughting, and the Centre of Effort of the Sails. Also, Weights and Sizes of Ropes; Masting, Rigging, and Sails of Steam Vessels, &c., &c. Tenth Edition, enlarged, with an Appendix. By ROBERT KIPPING, N.A., Sailmaker, Quayside, Newcastle. Illustrated. 2s. 6d.

155. *THE ENGINEER'S GUIDE TO THE ROYAL AND* MERCANTILE NAVIES. By a PRACTICAL ENGINEER. Revised by D. F. M'CARTHY, late of the Ordnance Survey Office, Southampton. 3s.

PHYSICAL SCIENCE, NATURAL PHILO-SOPHY, ETC.

1. *CHEMISTRY*, for the Use of Beginners. By Professor GEORGE FOWNES, F.R.S. With an Appendix, on the Application of Chemistry to Agriculture. 1s.

2. *NATURAL PHILOSOPHY*, Introduction to the Study of; for the Use of Beginners. By C. TOMLINSON, Lecturer on Natural Science in King's College School, London. Woodcuts. 1s. 6d.

4. *MINERALOGY*, Rudiments of; a concise View of the Properties of Minerals. By A. RAMSAY, Jun. Woodcuts and Steel Plates. 3s.

6. *MECHANICS*, Rudimentary Treatise on; being a concise Exposition of the General Principles of Mechanical Science, and their Applications. By CHARLES TOMLINSON, Lecturer on Natural Science in King's College School, London. Illustrated. 1s. 6d.

7. *ELECTRICITY;* showing the General Principles of Electrical Science, and the purposes to which it has been applied. By Sir W. SNOW HARRIS, F.R.S., &c. With considerable Additions by R. SABINE, C.E., F.S.A. Woodcuts. 1s. 6d.

7*. *GALVANISM*, Rudimentary Treatise on, and the General Principles of Animal and Voltaic Electricity. By Sir W. SNOW HARRIS. New Edition, revised, with considerable Additions, by ROBERT SABINE, C.E., F.S.A. Woodcuts. 1s. 6d.

8. *MAGNETISM;* being a concise Exposition of the General Principles of Magnetical Science, and the Purposes to which it has been applied. By Sir W. SNOW HARRIS. New Edition, revised and enlarged by H. M. NOAD, Ph.D., Vice-President of the Chemical Society, Author of "A Manual of Electricity," &c., &c. With 165 Wooocuts. 3s. 6d.

1. *THE ELECTRIC TELEGRAPH;* its History and Progress; with Descriptions of some of the Apparatus. By R. SABINE, C.E., F.S.A., &c. Woodcuts. 3s.

12. *PNEUMATICS*, for the Use of Beginners. By CHARLES TOMLINSON. Illustrated. 1s. 6d.

72. *MANUAL OF THE MOLLUSCA;* a Treatise on Recent and Fossil Shells. By Dr. S. P. WOODWARD, A.L.S. With Appendix by RALPH TATE, A.L.S., F.G.S. With numerous Plates and 300 Woodcuts, 6s. 6d. Cloth boards, 7s. 6d.

79**. *PHOTOGRAPHY*, Popular Treatise on; with a Description of the Stereoscope, &c. Translated from the French of D. Van Monckhoven, by W. H. Thornthwaite, Ph.D. Woodcuts. 1s. 6d.

96. *ASTRONOMY*. By the Rev. R. Main, M.A., F.R.S., &c. New and enlarged Edition, with an Appendix on "Spectrum Analysis." Woodcuts. 1s. 6d.

97. *STATICS AND DYNAMICS*, the Principles and Practice of; embracing also a clear development of Hydrostatics, Hydrodynamics, and Central Forces. By T. Baker, C.E. 1s. 6d.

138. *TELEGRAPH*, Handbook of the; a Manual of Telegraphy, Telegraph Clerks' Remembrancer, and Guide to Candidates for Employment in the Telegraph Service. By R. Bond. Fourth Edition, revised and enlarged: to which is appended, QUESTIONS on MAGNETISM, ELECTRICITY, and PRACTICAL TELEGRAPHY, for the Use of Students, by W. McGregor, First Assistant Superintendent, Indian Gov. Telegraphs. Woodcuts. 3s.

143. *EXPERIMENTAL ESSAYS.* By Charles Tomlinson. I. On the Motions of Camphor on Water. II. On the Motion of Camphor towards the Light. III. History of the Modern Theory of Dew. Woodcuts. 1s.

173. *PHYSICAL GEOLOGY*, partly based on Major-General Portlock's "Rudiments of Geology." By Ralph Tate, A.L.S., &c. Numerous Woodcuts. 2s.

174. *HISTORICAL GEOLOGY*, partly based on Major-General Portlock's "Rudiments." By Ralph Tate, A.L.S., &c. Woodcuts. 2s. 6d.

173 & 174. *RUDIMENTARY TREATISE ON GEOLOGY*, Physical and Historical. Partly based on Major-General Portlock's "Rudiments of Geology." By Ralph Tate, A.L.S., F.G.S., &c., &c. Numerous Illustrations. In One Volume. 4s. 6d.

183 & 184. *ANIMAL PHYSICS*, Handbook of. By Dionysius Lardner, D.C.L., formerly Professor of Natural Philosophy and Astronomy in University College, London. With 520 Illustrations. In One Volume, cloth boards. 7s. 6d.

⁎ *Sold also in Two Parts, as follows :—*

183. Animal Physics. By Dr. Lardner. Part I., Chapter I—VII. 4s.
184. Animal Physics. By Dr. Lardner. Part II. Chapter VIII—XVIII. 3s.

MINING, METALLURGY, ETC.

117. *SUBTERRANEOUS SURVEYING*, Elementary and Practical Treatise on, with and without the Magnetic Needle. By Thomas Fenwick, Surveyor of Mines, and Thomas Baker, C.E. Illustrated. 2s. 6d.

133. *METALLURGY OF COPPER ;* an Introduction to the Methods of Seeking, Mining, and Assaying Copper, and Manufacturing its Alloys. By Robert H. Lamborn, Ph.D. Woodcuts. 2s. 6d.

134. *METALLURGY OF SILVER AND LEAD.* A Description of the Ores; their Assay and Treatment, and valuable Constituents. By Dr. R. H. Lamborn. Woodcuts. 2s

135. *ELECTRO-METALLURGY;* Practically Treated. By Alexander Watt, F.R.S.S.A. New Edition, enlarged. Woodcuts. 2s. 6d.

172. *MINING TOOLS*, Manual of. For the Use of Mine Managers, Agents, Students, &c. Comprising Observations on the Materials from, and Processes by, which they are manufactured; their Special Uses, Applications, Qualities, and Efficiency. By William Morgans, Lecturer on Mining at the Bristol School of Mines. 2s. 6d.

172*. *MINING TOOLS, ATLAS* of Engravings to Illustrate the above, containing 235 Illustrations of Mining Tools, drawn to Scale. 4to. 4s. 6d.

Mining, Metallurgy, etc., *continued*.

176. *METALLURGY OF IRON*, a Treatise on the.   Containing Outlines of the History of Iron Manufacture, Methods of Assay, and Analyses of Iron Ores, Processes of Manufacture of Iron and Steel, &c.   By H. BAUERMAN, F.G.S., Associate of the Royal School of Mines.   Fourth Edition, revised and enlarged, with numerous Illustrations.   4s. 6d.

180. *COAL AND COAL MINING:* A Rudimentary Treatise on. By WARINGTON W. SMYTH, M.A., F.R.S., &c., Chief Inspector of the Mines of the Crown and of the Duchy of Cornwall.   New Edition, revised and corrected.   With numerous Illustrations.   3s. 6d.

EMIGRATION.

154. *GENERAL HINTS TO EMIGRANTS.* Containing Notices of the various Fields for Emigration.   With Hints on Preparation for Emigrating, Outfits, &c., &c.   With Directions and Recipes useful to the Emigrant.   With a Map of the World.   2s.

157. *THE EMIGRANT'S GUIDE TO NATAL.*   By ROBERT JAMES MANN, F.R.A.S., F.M.S.   Second Edition, carefully corrected to the present Date.   Map.   2s.

159. *THE EMIGRANT'S GUIDE TO AUSTRALIA, New South Wales, Western Australia, South Australia, Victoria, and Queensland.*   By the Rev. JAMES BAIRD, B.A.   Map.   2s. 6d.

160. *THE EMIGRANT'S GUIDE TO TASMANIA and NEW ZEALAND.*   By the Rev. JAMES BAIRD, B.A.   With a Map.   2s.

159 & *THE EMIGRANT'S GUIDE TO AUSTRALASIA.*   By the 160. Rev. J. BAIRD, B.A.   Comprising the above two volumes, 12mo, cloth boards. With Maps of Australia and New Zealand.   5s.

AGRICULTURE.

29. *THE DRAINAGE OF DISTRICTS AND LANDS.*   By G. DRYSDALE DEMPSEY, C.E.   Illustrated.   1s. 6d.
  ⁎ With "*Drainage of Towns and Buildings*," in One Vol., 3s. 6d.

63. *AGRICULTURAL ENGINEERING:* Farm Buildings, Motive Powers and Machinery of the Steading, Field Machines, and Implements. By G. H. ANDREWS, C.E.   Illustrated.   3s.

66. *CLAY LANDS AND LOAMY SOILS.*   By Professor DONALDSON.   1s.

131. *MILLER'S, MERCHANT'S, AND FARMER'S READY RECKONER*, for ascertaining at sight the value of any quantity of Corn, from One Bushel to One Hundred Quarters, at any given price, from £1 to £5 per quarter.   Together with the approximate values of Millstones and Millwork, &c.   1s.

140. *SOILS, MANURES, AND CROPS.*   (Vol. 1. OUTLINES OF MODERN FARMING.)   By R. SCOTT BURN.   Woodcuts.   2s.

141. *FARMING AND FARMING ECONOMY*, Notes, Historical and Practical, on.   (Vol. 2. OUTLINES OF MODERN FARMING.)   By R. SCOTT BURN.   Woodcuts.   3s.

142. *STOCK; CATTLE, SHEEP, AND HORSES.*   (Vol. 3. OUTLINES OF MODERN FARMING.)   By R. SCOTT BURN.   Woodcuts.   2s. 6d.

145. *DAIRY, PIGS, AND POULTRY*, Management of the.   By R. SCOTT BURN.   With Notes on the Diseases of Stock.   (Vol. 4. OUTLINES OF MODERN FARMING.)   Woodcuts.   2s.

146. *UTILIZATION OF SEWAGE, IRRIGATION, AND RECLAMATION OF WASTE LAND.*   (Vol. 5. OUTLINES OF MODERN FARMING.)   By R. SCOTT BURN.   Woodcuts.   2s. 6d.
  ⁎ Nos. 140-1-2-5-6, in One Vol., handsomely half-bound, entitled "OUTLINES OF MODERN FARMING."   By ROBERT SCOTT BURN.   Price 12s.

177. *FRUIT TREES*, The Scientific and Profitable Culture of.   From the French of DU BREUIL, Revised by GEO. GLENNY.   187 Woodcuts.   3s. 6d.

LONDON : CROSBY LOCKWOOD AND CO.,

FINE ARTS.

20. *PERSPECTIVE FOR BEGINNERS.* Adapted to Young Students and Amateurs in Architecture, Painting, &c. By GEORGE PYNE, Artist. Woodcuts. 2s.

40 & 41. *GLASS STAINING;* or, Painting on Glass, The Art of. Comprising Directions for Preparing the Pigments and Fluxes, laying them upon the Glass, and Firing or Burning in the Colours. From the German of Dr. GESSERT. To which is added, an Appendix on THE ART OF ENAMELLING, &c., with THE ART OF PAINTING ON GLASS. From the German of EMANUEL OTTO FROMBERG. In One Volume. 2s. 6d.

69. *MUSIC,* A Rudimentary and Practical Treatise on. With numerous Examples. By CHARLES CHILD SPENCER. 2s. 6d.

71. *PIANOFORTE,* The Art of Playing the. With numerous Exercises and Lessons. Written and Selected from the Best Masters, by CHARLES CHILD SPENCER. 1s. 6d.

181. *PAINTING POPULARLY EXPLAINED,* including Fresco, Oil, Mosaic, Water Colour, Water-Glass, Tempera, Encaustic, Miniature, Painting on Ivory, Vellum, Pottery, Enamel, Glass, &c. With Historical Sketches of the Progress of the Art by THOMAS JOHN GULLICK, assisted by JOHN TIMBS, F.S.A. Third Edition, revised and enlarged, with Frontispiece and Vignette. 5s.

186. *A GRAMMAR OF COLOURING,* applied to Decorative Painting and the Arts. By GEORGE FIELD. New Edition, enlarged and adapted to the Use of the Ornamental Painter and Designer. By ELLIS A. DAVIDSON, Author of "Drawing for Carpenters," &c. With two new Coloured Diagrams and numerous Engravings on Wood. 2s. 6d.

ARITHMETIC, GEOMETRY, MATHEMATICS, ETC.

32. *MATHEMATICAL INSTRUMENTS,* a Treatise on; in which their Construction and the Methods of Testing, Adjusting, and Using them are concisely Explained. By J. F. HEATHER, M.A., of the Royal Military Academy, Woolwich. Original Edition, in 1 vol., Illustrated. 1s. 6d.

** *In ordering the above, be careful to say, "Original Edition," or give the number in the Series (32) to distinguish it from the Enlarged Edition in 3 vols. (Nos. 168-9-70.)*

60. *LAND AND ENGINEERING SURVEYING,* a Treatise on; with all the Modern Improvements. Arranged for the Use of Schools and Private Students; also for Practical Land Surveyors and Engineers. By T. BAKER, C.E. New Edition, revised by EDWARD NUGENT, C.E. Illustrated with Plates and Diagrams. 2s.

61*. *READY RECKONER FOR THE ADMEASUREMENT OF LAND.* By ABRAHAM ARMAN, Schoolmaster, Thurleigh, Beds. To which is added a Table, showing the Price of Work, from 2s. 6d. to £1 per acre, and Tables for the Valuation of Land, from 1s. to £1,000 per acre, and from one pole to two thousand acres in extent, &c., &c. 1s. 6d.

76. *DESCRIPTIVE GEOMETRY,* an Elementary Treatise on; with a Theory of Shadows and of Perspective, extracted from the French of G. MONGE. To which is added, a description of the Principles and Practice of Isometrical Projection; the whole being intended as an introduction to the Application of Descriptive Geometry to various branches of the Arts. By J. F. HEATHER, M.A. Illustrated with 14 Plates. 2s.

178. *PRACTICAL PLANE GEOMETRY:* giving the Simplest Modes of Constructing Figures contained in one Plane and Geometrical Construction of the Ground. By J. F. HEATHER, M.A. With 215 Woodcuts. 2s.

179. *PROJECTION:* Orthographic, Topographic, and Perspective: giving the various Modes of Delineating Solid Forms by Constructions on a Single Plane Surface. By J. F. HEATHER, M.A. [*In preparation.*

** *The above three volumes will form a* COMPLETE ELEMENTARY COURSE OF MATHEMATICAL DRAWING.

Arithmetic, Geometry, Mathematics, etc., *continued.*

83. *COMMERCIAL BOOK-KEEPING.* With Commercial Phrases and Forms in English, French, Italian, and German. By JAMES HADDON, M.A., Arithmetical Master of King's College School, London. 1s.

84. *ARITHMETIC,* a Rudimentary Treatise on : with full Explanations of its Theoretical Principles, and numerous Examples for Practice. For the Use of Schools and for Self-Instruction. By J. R. YOUNG, late Professor of Mathematics in Belfast College. New Edition, with Index. 1s. 6d.

85.* A KEY to the above, containing Solutions in full to the Exercises, together with Comments, Explanations, and Improved Processes, for the Use of Teachers and Unassisted Learners. By J. R. YOUNG. 1s. 6d.

85. *EQUATIONAL ARITHMETIC,* applied to Questions of Interest,
85*. Annuities, Life Assurance, and General Commerce ; with various Tables by which all Calculations may be greatly facilitated. By W. HIPSLEY. 2s.

86. *ALGEBRA,* the Elements of. By JAMES HADDON, M.A., Second Mathematical Master of King's College School. With Appendix, containing miscellaneous Investigations, and a Collection of Problems in various parts of Algebra. 2s.

86* A KEY AND COMPANION to the above Book, forming an extensive repository of Solved Examples and Problems in Illustration of the various Expedients necessary in Algebraical Operations. Especially adapted for Self-Instruction. By J. R. YOUNG. 1s. 6d.

88. *EUCLID,* THE ELEMENTS OF : with many additional Propositions
89. and Explanatory Notes : to which is prefixed, an Introductory Essay on Logic. By HENRY LAW, C.E. 2s. 6d.

*** Sold also separately, viz. :—*

88. EUCLID, The First Three Books. By HENRY LAW, C.E. 1s.
89. EUCLID, Books 4, 5, 6, 11, 12. By HENRY LAW, C.E. 1s. 6d.

90. *ANALYTICAL GEOMETRY AND CONIC SECTIONS,* a Rudimentary Treatise on. By JAMES HANN, late Mathematical Master of King's College School, London. A New Edition, re-written and enlarged by J. R. YOUNG, formerly Professor of Mathematics at Belfast College. 2s.

91. *PLANE TRIGONOMETRY,* the Elements of. By JAMES HANN, formerly Mathematical Master of King's College, London. 1s.

92. *SPHERICAL TRIGONOMETRY,* the Elements of. By JAMES HANN. Revised by CHARLES H. DOWLING, C.E. 1s.
*** Or with " The Elements of Plane Trigonometry," in One Volume, 2s.*

93. *MENSURATION AND MEASURING,* for Students and Practical Use. With the Mensuration and Levelling of Land for the Purposes of Modern Engineering. By T. BAKER, C.E. New Edition, with Corrections and Additions by E. NUGENT, C.E. Illustrated. 1s. 6d.

94. *LOGARITHMS,* a Treatise on ; with Mathematical Tables for facilitating Astronomical, Nautical, Trigonometrical, and Logarithmic Calculations ; Tables of Natural Sines and Tangents and Natural Cosines. By HENRY LAW, C.E. Illustrated. 2s. 6d.

101*. *MEASURES, WEIGHTS, AND MONEYS OF ALL NATIONS,* and an Analysis of the Christian, Hebrew, and Mahometan Calendars. By W. S. B. WOOLHOUSE, F.R.A.S., &c. 1s. 6d.

102. *INTEGRAL CALCULUS,* Rudimentary Treatise on the. By HOMERSHAM COX, B.A. Illustrated. 1s.

103. *INTEGRAL CALCULUS,* Examples on the. By JAMES HANN, late of King's College, London. Illustrated. 1s.

101. *DIFFERENTIAL CALCULUS,* Examples of the. By W. S. B. WOOLHOUSE, F.R.A.S., &c. 1s. 6d.

104. *DIFFERENTIAL CALCULUS,* Examples and Solutions of the. By JAMES HADDON, M.A. 1s.

LONDON : CROSBY LOCKWOOD AND CO.,

Arithmetic, Geometry, Mathematics, etc., *continued.*

105. *MNEMONICAL LESSONS.* — GEOMETRY, ALGEBRA, AND TRIGONOMETRY, in Easy Mnemonical Lessons. By the Rev. THOMAS PENYNGTON KIRKMAN, M.A. 1s. 6d.

136. *ARITHMETIC,* Rudimentary, for the Use of Schools and Self-Instruction. By JAMES HADDON, M.A. Revised by ABRAHAM ARMAN. 1s. 6d.

137. A KEY to HADDON's RUDIMENTARY ARITHMETIC. By A. ARMAN. 1s. 6d.

147. *ARITHMETIC,* STEPPING-STONE TO; being a Complete Course of Exercises in the First Four Rules (Simple and Compound), on an entirely new principle. For the Use of Elementary Schools of every Grade. Intended as an Introduction to the more extended works on Arithmetic. By ABRAHAM ARMAN. 1s.

148. A KEY TO STEPPING-STONE TO ARITHMETIC. By A. ARMAN. 1s.

158. *THE SLIDE RULE, AND HOW TO USE IT;* containing full, easy, and simple Instructions to perform all Business Calculations with unexampled rapidity and accuracy. By CHARLES HOARE, C.E. With a Slide Rule in tuck of cover. 3s.

168. *DRAWING AND MEASURING INSTRUMENTS.* Including—I. Instruments employed in Geometrical and Mechanical Drawing, and in the Construction, Copying, and Measurement of Maps and Plans. II. Instruments used for the purposes of Accurate Measurement, and for Arithmetical Computations. By J. F. HEATHER, M.A., late of the Royal Military Academy, Woolwich, Author of "Descriptive Geometry," &c., &c. Illustrated. 1s. 6d.

169. *OPTICAL INSTRUMENTS.* Including (more especially) Telescopes, Microscopes, and Apparatus for producing copies of Maps and Plans by Photography. By J. F. HEATHER, M.A. Illustrated. 1s. 6d.

170. *SURVEYING AND ASTRONOMICAL INSTRUMENTS.* Including—I. Instruments Used for Determining the Geometrical Features of a portion of Ground. II. Instruments Employed in Astronomical Observations. By J. F. HEATHER, M.A. Illustrated. 1s. 6d.

*** *The above three volumes form an enlargement of the Author's original work, "Mathematical Instruments: their Construction, Adjustment, Testing, and Use," the Eleventh Edition of which is on sale, price 1s. 6d. (See No. 32 in the Series.)*

168.⎫
169.⎬ *MATHEMATICAL INSTRUMENTS.* By J. F. HEATHER, M.A. Enlarged Edition, for the most part entirely re-written. The 3 Parts as
170.⎭ above, in One thick Volume. With numerous Illustrations. 4s. 6d.

185. *THE COMPLETE MEASURER;* setting forth the Measurement of Boards, Glass, &c., &c.; Unequal-sided, Square-sided, Octagonal-sided, Round Timber and Stone, and Standing Timber. With just Allowances for the Bark in the respective species of Trees, and proper deductions for the waste in hewing the trees, &c.; also a Table showing the solidity of hewn or eight-sided timber, or of any octagonal-sided column. Compiled for the accommodation of Timber-growers, Merchants, and Surveyors, Stonemasons, Architects, and others. By RICHARD HORTON. Third Edition, with considerable and valuable additions. 4s. [*Just published.*

LEGAL TREATISES.

50. *THE LAW OF CONTRACTS FOR WORKS AND SER-VICES.* By DAVID GIBBONS. Third Edition, revised and considerably enlarged. 3s. [*Just published.*

151. *A HANDY BOOK ON THE LAW OF FRIENDLY, IN-DUSTRIAL & PROVIDENT BUILDING & LOAN SOCIETIES.* With copious Notes. By NATHANIEL WHITE, of H.M. Civil Service. 1s.

163. *THE LAW OF PATENTS FOR INVENTIONS;* and on the Protection of Designs and Trade Marks. By F. W. CAMPIN, Barrister-at-Law. 2s.

MISCELLANEOUS VOLUMES.

36. *A DICTIONARY OF TERMS used in ARCHITECTURE, BUILDING, ENGINEERING, MINING, METALLURGY, ARCHÆOLOGY, the FINE ARTS, &c.* By JOHN WEALE. Fifth Edition. Revised by ROBERT HUNT, F.R.S., Keeper of Mining Records. Numerous Illustrations. 5s.

112. *MANUAL OF DOMESTIC MEDICINE.* By R. GOODING, B.A., M.B. Intended as a Family Guide in all Cases of Accident and Emergency. 2s.

112*. *MANAGEMENT OF HEALTH.* A Manual of Home and Personal Hygiene. By the Rev. JAMES BAIRD, B.A. 1s.

113. *FIELD ARTILLERY ON SERVICE.* By TAUBERT, Captain Prussian Artillery. Translated by Lieut.-Col. H. H. MAXWELL. 1s. 6d.

113*. *SWORDS, AND OTHER ARMS.* By Col. MAREY. Translated by Col. H. H. MAXWELL. With Plates. 1s.

150. *LOGIC*, Pure and Applied. By S. H. EMMENS. Third Edition. 1s. 6d.

152. *PRACTICAL HINTS FOR INVESTING MONEY.* With an Explanation of the Mode of Transacting Business on the Stock Exchange. By FRANCIS PLAYFORD, Sworn Broker. 1s. 6d.

153. *SELECTIONS FROM LOCKE'S ESSAYS ON THE HUMAN UNDERSTANDING.* With Notes by S. H. EMMENS. 2s.

193. *HANDBOOK OF FIELD FORTIFICATION*, intended for the Guidance of Officers Preparing for Promotion, and especially adapted to the requirements of Beginners. By Major W. W. KNOLLYS, F.R.G.S., 93rd Sutherland Highlanders, &c. With 163 Woodcuts. 3s.

194. *THE HOUSEHOLD AND ITS MANAGEMENT:* Being a Guide to Housekeeping, Practical Cookery, Pickling and Preserving, Household Work, Dairy Management, the Table and Dessert, Cellarage of Wines, Home-brewing and Wine-making, the Boudoir and Dressing-room, Travelling, Stable Economy, Gardening Operations, &c. By AN OLD HOUSEKEEPER. 3s. 6d.

EDUCATIONAL AND CLASSICAL SERIES.

HISTORY.

1. **England, Outlines of the History of;** more especially with reference to the Origin and Progress of the English Constitution. A Text Book for Schools and Colleges. By WILLIAM DOUGLAS HAMILTON, F.S.A., of Her Majesty's Public Record Office. Fourth Edition, revised. Maps and Woodcuts. 5s.; cloth boards, 6s.

5. **Greece, Outlines of the History of;** in connection with the Rise of the Arts and Civilization in Europe. By W. DOUGLAS HAMILTON, of University College, London, and EDWARD LEVIEN, M.A., of Balliol College, Oxford. 2s. 6d.; cloth boards, 3s. 6d.

7. **Rome, Outlines of the History of:** from the Earliest Period to the Christian Era and the Commencement of the Decline of the Empire. By EDWARD LEVIEN, of Balliol College, Oxford. Map, 2s. 6d.; cl. bds. 3s. 6d.

9. **Chronology of History, Art, Literature, and Progress,** from the Creation of the World to the Conclusion of the Franco-German War. The Continuation by W. D. HAMILTON, F.S.A., of Her Majesty's Record Office. 3s.; cloth boards, 3s. 6d.

50. **Dates and Events in English History,** for the use of Candidates in Public and Private Examinations. By the Rev. E. RAND. 1s.

ENGLISH LANGUAGE AND MISCELLANEOUS.

11. **Grammar of the English Tongue,** Spoken and Written. With an Introduction to the Study of Comparative Philology. By HYDE CLARKE, D.C.L. Third Edition. 1s.

11*. **Philology:** Handbook of the Comparative Philology of English, Anglo-Saxon, Frisian, Flemish or Dutch, Low or Platt Dutch, High Dutch or German, Danish, Swedish, Icelandic, Latin, Italian, French, Spanish, and Portuguese Tongues. By HYDE CLARKE, D.C.L. 1s.

12. **Dictionary of the English Language,** as Spoken and Written. Containing above 100,000 Words. By HYDE CLARKE, D.C.L. 3s. 6d.; cloth boards, 4s. 6d.; complete with the GRAMMAR, cloth bds., 5s. 6d.

48. **Composition and Punctuation,** familiarly Explained for those who have neglected the Study of Grammar. By JUSTIN BRENAN. 16th Edition. 1s.

49. **Derivative Spelling-Book:** Giving the Origin of Every Word from the Greek, Latin, Saxon, German, Teutonic, Dutch, French, Spanish, and other Languages; with their present Acceptation and Pronunciation. By J. ROWBOTHAM, F.R.A.S. Improved Edition. 1s. 6d.

51. **The Art of Extempore Speaking:** Hints for the Pulpit, the Senate, and the Bar. By M. BAUTAIN, Vicar-General and Professor at the Sorbonne. Translated from the French. Sixth Edition, carefully corrected. 2s. 6d.

52. **Mining and Quarrying,** with the Sciences connected therewith. First Book of, for Schools. By J. H. COLLINS, F.G.S., Lecturer to the Miners' Association of Cornwall and Devon. 1s.

53. **Places and Facts in Political and Physical Geography,** for Candidates in Public and Private Examinations. By the Rev. EDGAR RAND, B.A. 1s.

54. **Analytical Chemistry,** Qualitative and Quantitative, a Course of. To which is prefixed, a Brief Treatise upon Modern Chemical Nomenclature and Notation. By WM. W. PINK, Practical Chemist, &c., and GEORGE E. WEBSTER, Lecturer on Metallurgy and the Applied Sciences, Nottingham. 2s.

THE SCHOOL MANAGERS' SERIES OF READING BOOKS,

Adapted to the Requirements of the New Code. Edited by the Rev. A. R. GRANT, Rector of Hitcham, and Honorary Canon of Ely; formerly H.M. Inspector of Schools.

INTRODUCTORY PRIMER, 3d.

	s.	d.			s.	d.
FIRST STANDARD	0	6	FOURTH STANDARD		1	2
SECOND ,,	0	10	FIFTH ,,		1	6
THIRD ,,	1	0	SIXTH ,,		1	6

LESSONS FROM THE BIBLE. Part I. Old Testament. 1s.
LESSONS FROM THE BIBLE. Part II. New Testament, to which is added THE GEOGRAPHY OF THE BIBLE, for very young Children. By Rev. C. THORNTON FORSTER. 1s. 2d. *.* Or the Two Parts in One Volume. 2s.

FRENCH.

24. **French Grammar.** With Complete and Concise Rules on the Genders of French Nouns. By G. L. STRAUSS, Ph.D. 1s.

25. **French-English Dictionary.** Comprising a large number of New Terms used in Engineering, Mining, on Railways, &c. By ALFRED ELWES. 1s. 6d.

26. **English-French Dictionary.** By ALFRED ELWES. 2s.

25,26. **French Dictionary** (as above). Complete, in One Vol., 3s.; cloth boards, 3s. 6d. *.* Or with the GRAMMAR, cloth boards, 4s. 6d.

French, *continued.*

47. **French and English Phrase Book :** containing Introductory Lessons, with Translations, for the convenience of Students ; several Vocabularies of Words, a Collection of suitable Phrases, and Easy Familiar Dialogues. 1s.

GERMAN.

39. **German Grammar.** Adapted for English Students, from Heyse's Theoretical and Practical Grammar, by Dr. G. L. STRAUSS. 1s.

40. **German Reader :** A Series of Extracts, carefully culled from the most approved Authors of Germany ; with Notes, Philological and Explanatory. By G. L. STRAUSS, Ph.D. 1s.

41. **German Triglot Dictionary.** By NICHOLAS ESTERHAZY, S. A. HAMILTON. Part I. English-German-French. 1s.

42. **German Triglot Dictionary.** Part II. German-French-English. 1s.

43. **German Triglot Dictionary.** Part III. French-German-English. 1s.

41-43. German Triglot Dictionary (as above), in One Vol., 3s. ; cloth boards, 4s. *₊* Or with the GERMAN GRAMMAR, cloth boards, 5s.

ITALIAN.

27. **Italian Grammar,** arranged in Twenty Lessons, with a Course of Exercises. By ALFRED ELWES. 1s.

28. **Italian Triglot Dictionary,** wherein the Genders of all the Italian and French Nouns are carefully noted down. By ALFRED ELWES. Vol. 1. Italian-English-French. 2s.

30. **Italian Triglot Dictionary.** By A. ELWES. Vol. 2. English-French-Italian. 2s.

32. **Italian Triglot Dictionary.** By ALFRED ELWES. Vol. 3. French-Italian-English. 2s.

28,30, Italian Triglot Dictionary (as above). In One Vol., 6s. ;
32.   cloth boards, 7s. 6d. *₊* Or with the ITALIAN GRAMMAR, cloth bds., 8s. 6d.

SPANISH AND PORTUGUESE.

34. **Spanish Grammar,** in a Simple and Practical Form. With a Course of Exercises. By ALFRED ELWES. 1s. 6d.

35. **Spanish-English and English-Spanish Dictionary.** Including a large number of Technical Terms used in Mining, Engineering, &c., with the proper Accents and the Gender of every Noun. By ALFRED ELWES. 4s. ; cloth boards, 5s. *₊* Or with the GRAMMAR, cloth boards, 6s.

55. **Portuguese Grammar,** in a Simple and Practical Form. With a Course of Exercises. By ALFRED ELWES, Author of " A Spanish Grammar," &c. 1s. 6d.   [*Just published*.

HEBREW.

46*. **Hebrew Grammar.** By Dr. BRESSLAU. 1s. 6d.

44. **Hebrew and English Dictionary,** Biblical and Rabbinical ; containing the Hebrew and Chaldee Roots of the Old Testament Post-Rabbinical Writings. By Dr. BRESSLAU. 6s. *₊* Or with the GRAMMAR, 7s.

46. **English and Hebrew Dictionary.** By Dr. BRESSLAU. 3s.

44,46. **Hebrew Dictionary** (as above), in Two Vols., complete, with
46*.   the GRAMMAR, cloth boards, 12s.

LATIN.

19. **Latin Grammar.** Containing the Inflections and Elementary Principles of Translation and Construction. By the Rev. THOMAS GOODWIN, M.A., Head Master of the Greenwich Proprietary School. 1s.

20. **Latin-English Dictionary.** Compiled from the best Authorities. By the Rev. THOMAS GOODWIN, M.A. 2s.

22. **English-Latin Dictionary;** together with an Appendix of French and Italian Words which have their origin from the Latin. By the Rev. THOMAS GOODWIN, M.A. 1s. 6d.

20,22. **Latin Dictionary** (as above). Complete in One Vol., 3s. 6d.; cloth boards, 4s. 6d. *⁎* Or with the GRAMMAR, cloth boards, 5s. 6d.

LATIN CLASSICS. With Explanatory Notes in English.

1. **Latin Delectus.** Containing Extracts from Classical Authors, with Genealogical Vocabularies and Explanatory Notes, by HENRY YOUNG, lately Second Master of the Royal Grammar School, Guildford. 1s.

2. **Cæsaris** Commentarii de Bello Gallico. Notes, and a Geographical Register for the Use of Schools, by H. YOUNG. 2s.⁄

12. **Ciceronis** Oratio pro Sexto Roscio Amerino. Edited, with an Introduction, Analysis, and Notes Explanatory and Critical, by the Rev. JAMES DAVIES, M.A. 1s.

14. **Ciceronis** Cato Major, Lælius, Brutus, sive de Senectute, de Amicitia, de Claris Oratoribus Dialogi. With Notes by W. BROWNRIGG SMITH, M.A., F.R.G.S. 2s.

3. **Cornelius Nepos.** With Notes. Intended for the Use of Schools. By H. YOUNG. 1s.

6. **Horace;** Odes, Epode, and Carmen Sæculare. Notes by H. YOUNG. 1s. 6d.

7. **Horace;** Satires, Epistles, and Ars Poetica. Notes by W. BROWN-RIGG SMITH, M.A., F.R.G.S. 1s. 6d.

21. **Juvenalis** Satiræ. With Prolegomena and Notes by T. H. S. ESCOTT, B.A., Lecturer on Logic at King's College, London. 1s. 6d.

16. **Livy:** History of Rome. Notes by H. YOUNG and W. B. SMITH, M.A. Part 1. Books i., ii., 1s. 6d.

16*. ——— Part 2. Books iii., iv., v., 1s. 6d.

17. ——— Part 3 Books xxi., xxii., 1s. 6d.

8. **Sallustii** Crispi Catalina et Bellum Jugurthinum. Notes Critical and Explanatory, by W. M. DONNE, B.A., Trinity College, Cambridge. 1s. 6d.

10. **Terentii** Adelphi, Hecyra, Phormio. Edited, with Notes, Critical and Explanatory, by the Rev. JAMES DAVIES, M.A. 2s.

9. **Terentii** Andria et Heautontimorumenos. With Notes, Critical and Explanatory, by the Rev. JAMES DAVIES, M.A. 1s. 6d.

11. **Terentii** Eunuchus, Comœdia. Edited, with Notes, by the Rev. JAMES DAVIES, M.A. 1s. 6d. Or the Adelphi, Andria, and Eunuchus, 3 vols. in 1, cloth boards, 6s.

4. **Virgilii** Maronis Bucolica et Georgica. With Notes on the Bucolics by W. RUSHTON, M.A., and on the Georgics by H. YOUNG. 1s. 6d.

5. **Virgilii** Maronis Æneis. Notes, Critical and Explanatory, by H. YOUNG. 2s.

19. **Latin Verse Selections,** from Catullus, Tibullus, Propertius, and Ovid. Notes by W. B. DONNE, M.A., Trinity College, Cambridge. 2s.

20. **Latin Prose Selections,** from Varro, Columella, Vitruvius, Seneca, Quintilian, Florus, Velleius Paterculus, Valerius Maximus Suetonius, Apuleius, &c. Notes by W. B. DONNE, M.A. 2s.

☞ *Other Volumes are in Preparation.*

GREEK.

14. **Greek Grammar,** in accordance with the Principles and Philological Researches of the most eminent Scholars of our own day. By HANS CLAUDE HAMILTON. 1s. 6d.

15,17. **Greek Lexicon.** Containing all the Words.in General Use, with their Significations, Inflections, and Doubtful Quantities. By HENRY R. HAMILTON. Vol. 1. Greek-English, 2s.; Vol. 2. English-Greek, 2s. Or the Two Vols. in One, 4s.: cloth boards, 5s.

14,15. **Greek Lexicon** (as above). Complete, with the GRAMMAR, in
17.   One Vol., cloth boards, 6s.

GREEK CLASSICS. With Explanatory Notes in English.

1. **Greek Delectus.** Containing Extracts from Classical Authors, with Genealogical Vocabularies and Explanatory Notes, by H. YOUNG. New Edition, with an improved and enlarged Supplementary Vocabulary, by JOHN HUTCHISON, M.A., of the High School, Glasgow. 1s.

30. **Æschylus:** Prometheus Vinctus : The Prometheus Bound. From the Text of DINDORF. Edited, with English Notes, Critical and Explanatory, by the Rev. JAMES DAVIES, M.A. 1s.

32. **Æschylus:** Septem Contra Thebes : The Seven against Thebes. From the Text of DINDORF. Edited, with English Notes, Critical and Explanatory, by the Rev. JAMES DAVIES, M.A. 1s.

40. **Aristophanes:** Acharnians. Chiefly from the Text of C. H. WEISE. With Notes, by C. S. T. TOWNSHEND, M.A. 1s. 6d.

26. **Euripides:** Alcestis. Chiefly from the Text of DINDORF. With Notes, Critical and Explanatory, by JOHN MILNER, B.A. 1s.

23. **Euripides:** Hecuba and Medea. Chiefly from the Text of DINDORF. With Notes, Critical and Explanatory, by W. BROWNRIGG SMITH, M.A., F.R.G.S. 1s. 6d.

14-17. **Herodotus,** The History of, chiefly after the Text of GAISFORD. With Preliminary Observations and Appendices, and Notes, Critical and Explanatory, by T. H. L. LEARY, M.A., D.C.L.
      Part 1. Books i., ii. (The Clio and Euterpe), 2s.
      Part 2. Books iii., iv. (The Thalia and Melpomene), 2s.
      Part 3. Books v.-vii. (The Terpsichore, Erato, and Polymnia), 2s.
      Part 4. Books viii., ix. (The Urania and Calliope) and Index, 1s. 6d.

5-12. **Homer,** The Works of. According to the Text of BAEUMLEIN. With Notes, Critical and Explanatory, drawn from the best and latest Authorities, with Preliminary Observations and Appendices, by T. H. L. LEARY, M.A., D.C.L.

THE ILIAD : Part 1. Books i. to vi., 1s. 6d. | Part 3. Books xiii. to xviii., 1s. 6d.
           Part 2. Books vii. to xii., 1s. 6d. | Part 4. Books xix. to xxiv., 1s. 6d.

THE ODYSSEY: Part 1. Books i. to vi., 1s. 6d. | Part 3. Books xiii. to xviii., 1s. 6d.
           Part 2. Books vii. to xii., 1s. 6d. | Part 4. Books xix. to xxiv., and
                                    Hymns, 2s.

4. **Lucian's Select Dialogues.** The Text carefully revised, with Grammatical and Explanatory Notes, by H. YOUNG. 1s.

13. **Plato's Dialogues:** The Apology of Socrates, the Crito, and the Phædo. From the Text of C. F. HERMANN. Edited with Notes, Critical and Explanatory, by the Rev. JAMES DAVIES, M.A. 2s.

18. **Sophocles:** Œdipus Tyrannus. Notes by H. YOUNG. 1s.

20. **Sophocles:** Antigone. From the Text of DINDORF. Notes, Critical and Explanatory, by the Rev. JOHN MILNER, B.A. 2s.

41. **Thucydides:** History of the Peloponnesian War. Notes by H. YOUNG. Book 1. 1s.

2, 3. **Xenophon's Anabasis;** or, The Retreat of the Ten Thousand. Notes and a Geographical Register, by H. YOUNG. Part 1. Books i. to iii., 1s. Part 2. Books iv. to vii., 1s.

42. **Xenophon's Panegyric** on Agesilaus. Notes and Introduction by LL. F. W. JEWITT. 1s. 6d.

☞ *Other Volumes are in Preparation.*

CROSBY LOCKWOOD AND CO., 7, STATIONERS' HALL COURT, E.C.

MEASURES, WEIGHTS, AND MONEYS OF ALL NATIONS, and an Analysis of the Christian, Hebrew, and Mahometan Calendars. By W. S. B. Woolhouse, F.R.A.S., &c. 1s. 6d.

INTEGRAL CALCULUS, Rudimentary Treatise on the. By Homersham Cox, B.A. Illustrated. 1s.

INTEGRAL CALCULUS, Examples on the. By James Hann, late of King's College, London. Illustrated. 1s.

DIFFERENTIAL CALCULUS, Elements of the. By W. S. B. Woolhouse, F.R.A.S., &c. 1s. 6d.

DIFFERENTIAL CALCULUS, Examples and Solutions in the. By James Haddon, M.A. 1s.

GEOMETRY, ALGEBRA, and TRIGONOMETRY, in Easy Mnemonical Lessons. By the Rev. Thomas Penyngton Kirkman, M.A. 1s. 6d.

MILLER'S, MERCHANT'S, AND FARMER'S READY RECKONER, for ascertaining at sight the value of any quantity of Corn, from one Bushel to one hundred Quarters, at any given price, from £1 to £5 per quarter. Together with the approximate values of Millstones and Millwork, &c. 1s.

ARITHMETIC, Rudimentary, for the use of Schools and Self-Instruction. By James Haddon, M.A. Revised by Abraham Arman. 1s. 6d.

A KEY to Rudimentary Arithmetic. By A. Arman. 1s. 6d.

ARITHMETIC, Stepping-Stone to; being a complete course of Exercises in the First Four Rules (Simple and Compound), on an entirely new principle. For the Use of Elementary Schools of every Grade. Intended as an Introduction to the more extended works on Arithmetic. By Abraham Arman. 1s.

A KEY to Stepping-Stone to Arithmetic. By A. Arman. 1s.

THE SLIDE RULE, AND HOW TO USE IT, containing full, easy, and simple instructions to perform all Business Calculations with unexampled rapidity and accuracy. By Charles Hoare, C.E. With a Slide Rule in tuck of cover. 3s

STATICS AND DYNAMICS, the Principles and Practice of; with those of Liquids and Gases. By T. Baker, C.E. Second Edition, revised by E. Nugent, C.E. Many Illustrations. 1s. 6d.

CROSBY LOCKWOOD & CO., 7, STATIONERS' HALL COURT, E.